ABW-1501

W9-CCW-431

Digital Multimedia Cross-Industry Guide

DIGITAL MULTIMEDIA CROSS-INDUSTRY GUIDE

EDITOR

Philip V. W. Dodds

Managing Editor Lee E. Fitzpatrick

FOCAL PRESS
Boston Oxford Melbourne Singapore Toronto Munich New Delhi Tokyo

This book was acquired, developed, and produced by Manning Publications Co.

Copyediting: Ruth Lofgren
Typesetting: Lauralee Butler Reinke

Manning Publications Co.
3 Lewis Street
Greenwich, CT 06830

Focal Press is an imprint of Butterworth–Heinemann.

Copyright © 1995 by Butterworth–Heinemann.

A member of the Reed Elsevier group

Library of Congress Cataloging-in-Publication Data
Digital multimedia cross-industry guide / editor, Philip V. W. Dodds.
    p. cm.
   Includes index.
   ISBN 0-240-80205-5 (hardcover)
   1. Multimedia systems.   I. Dodds, Philip V. W.
QA75.575.D54 1995
384--dc20   95-1586
                                            CIP

British Library Cataloguing-in-Publication Data
A catalogue record for this book is available from the British Library.

The publisher offers discounts on bulk orders of this book.
For information, please write:

Manager of Special Sales
Butterworth–Heinemann
313 Washington Street
Newton, MA 02158–1626

10 9 8 7 6 5 4 3 2 1

Printed in the United States of America

Contents

v

7   Computer Software   179

Preface

Several years ago Marjan Bace of Manning Publications approached me with the idea of writing a book about multimedia. I rejected the idea out of hand. There are too many such books in production, I said. And besides, as soon as one is written about any of the technological components, it is already obsolete. Books on digital video, CD-ROMs, and multimedia computers are a dime a dozen and not really interesting—at least not to the strategists and decision makers who are shaping the future of technology. Nope, not interested, I reiterated.

Marjan listened patiently and then posed the simple question that ultimately defined the structure of this book: What do you think *would* be relevant? And to whom? I explained that what people call "multimedia" is really a process of adoption and adaptation occurring within a variety of industries at different rates in different degrees. The audience are members of all these industries, and they are all confused about each other.

Multimedia is in broadcast and cable television, video games, personal computers, and is carried over telecommunications networks. A lot of people think digital multimedia will converge these into one unified industry and market and that it's all imploding into soon-to-be ubiquitous computer appliances. After all, they say, bits are bits and can be computed anywhere. Others think that the conversion to digital media will allow them to penetrate new markets and successfully apply their old business models and rules. But, they're wrong; it's just *not* that simple.

Why is that, he asked? (Marjan often asks confoundingly simple questions that echo for months and months.) For one thing, I said, groping for an answer, few understand that the various multimedia industries have their own history and legacy of technologies. For another, there are lifetimes worth of experiences built up in narrow business and technical cultures that may not graft easily to other markets even though it appears that technology ought to transfer gracefully. Finally, multimedia dabblers have just entered a process of discovery about what's possible. In their excitement they have created new and often alluring visions. But many of these visions don't have a practical or financial foundation in the context of existing business and industries.

People in business need to understand the differences and sameness among the industry players if they are to make decisions, I asserted. Nearly two years later I am more convinced of this need than ever before. Hence this book.

This cross-industry study of multimedia describes the radically different "ecological" conditions within related industries while examining the historical roots for common links. Finding commonality between separate industries provides a fascinating pattern—a kind of genetic mapping of multimedia. Recognizing some of the market rules that have evolved over time provides a completely new and highly valuable strategic aid to the architects of twenty-first-century information systems.

The contributing authors have given us their own, unique perspective of their world, giving us a glimpse of the things they believe are important, relevant, and real to their businesses. I wish to thank each of them for their hard work, patience and support of this project. I have certainly learned a lot from each of them, and my views of their businesses will never be the same.

Predictably, I found that we have barely scratched the surface with this book. The diversity in each industrial ecosystem is vast. The forces for change are often non-technical, sometimes arbitrary and seemingly capricious. But this is as good a starting place as I can think of. Please don't throw me in that briar patch again, Marjan, my brain still hurts!

Philip V. W. Dodds
pdodds@ima.org

Contributors

Mr. Philip V. W. Dodds
IMA
48 Maryland Ave.
Suite 202
Annapolis, MD 21401-8011

The Business of Multimedia

Mr. Michael Rau
Senior Vice President
Science & Technology
National Association of
   Broadcasters
1771 N. Street, N.W.
Washington, D.C. 20036-2891

Broadcast Television

Dr. Walt Ciciora
45 Hulls Farm Road
Southport, CT 06490-1027

CABLE Television:
   Views on the Set-Top Box

Mr. Howard Mirowitz
Vice President
Advanced Product Planning
Mitsubishi Electronics America
5665 Plaza Drive
Cypress, CA 90630

Consumer Electronics

Mr. Gary M. Brinck
System Architect
IBM Multimedia Systems
P.O. Box 1328
MS 1508
Boca Raton, FL 33429-1328

Handheld Devices

Mr. Monty McGraw   *Computer Hardware*
Manager System Research
Compaq Computer Corporation
M090802
20555 State Highway #249
Houston, TX 77070

Mr. Jim Green   *Computer Software*
Program Manager
Microsoft Corporation
One Microsoft Way
Redmond, Washington 98052

Mr. Harry L. Bosco   *ATM Deployment:*
Vice President ATM Platform   *Architectures and Applications*
    Organization
AT&T Network Systems
Room 3B-401F
200 Schulz Drive
Red Bank, NJ 07701

Mr. Philippe Clarke   *Telecommunications*
MCI International, Inc.
1133 19th Street, NW, 5th Floor
Washington, DC 20036
202-416-5770

1

The Business of Multimedia

PHILIP V. W. DODDS

The computer industry's stunning advances in performance and cost reduction have created opportunities for digital technologies to invade new and previously unrelated industries. A wholesale shift is under way from analog media, such as composite television signals and vinyl records, to precisely calculated digital representations like high-definition TV and compact discs. Once in digital form, information is no longer captive to one industrial distribution system. In theory, the same digital information stream can be viewed at home, in a car, at your office, or anywhere, thus potentially uniting isolated information systems and industries.

These developments have spurred an interest in the remaking of traditional businesses as well as developing entirely new media-based markets. Now there is the potential to invade and take market share away from highly entrenched service providers. For these reasons, digital multimedia have become one of the glamour industries for the nineties, replete with publicity, hype, investments, and new business ventures. Once highly skeptical of multimedia, virtually all technology-based companies are forming new divisions, investing and partnering with apparently strange bedfellows in an attempt to formulate a stake in the new digital world order.

No Longer a Choice for Businesspeople

Whether or not multimedia will impact on most businesses is no longer the question it once was. For years, even within leading technology companies, many asked if this multimedia "thing" was ever really going to happen. After all, who really needs sound or video pictures in their desktop computer? And, why on earth would a telephone company care about interactivity? Within corporations, many training departments embraced the early versions of multimedia technology in the form of analog laser disc-based interactive video systems. Such technology has been viewed as a highly specialized case that remains too expensive and experimental to become part of the corporate information system strategy. Now that most computers are delivered "multimedia ready," multimedia technology has moved from being the exception to being very nearly the rule.

For some, the advent of interactive multimedia technology continues to be viewed with deep suspicion. The idea that sound and video might appear on computer screens looks to many like too much fun and, therefore, is not defensible in the workplace. No previous model for productivity has included television-like qualities and there is simply no place for fun and games in the workplace. As is often the case in the early stages of new markets, few envisioned the inevitable change in technological capability that has crept up on them. Fewer still understand the potential implications.

Predictably, the onset of multimedia-capable products has generated a spate of new conferences and exhibitions. Virtually every electronics-related event now has a multimedia pavilion accompanied by high marquee-value speakers who supposedly have the inside track on the next hot revolutionary step on the multimedia ladder. By preannouncing the existence of a relatively undefined market, these conferences deliver confusing and often contradictory hype with the hope that if someone announces a trend, it will develop. Many ruefully observe that the real money in the multimedia industry has been made in trade show management rather than with the technology.

If trade shows and conferences lack substance about the emerging multimedia industry, where do you turn for knowledge if your business appears to be affected? The dozens of thousand-dollar-per-year newsletters and magazines report snippets of late-breaking deals and partnerships, but the reporting rarely has sufficient depth on which to base serious investment decisions. Detailed market studies, often costing in the six-figure range, plot a maze of possible long-term statistics that usually confound rather than clarify. When it comes to new markets, there really is not any reliable source of knowledge until it has been established for some length of time, usually several years.

Multimedia Skill Sets

Many industries are claiming a major stake in multimedia. Few share the same vision for how it will be implemented. All tend to depict their own industry as the center from which the multimedia wave will emanate; nearly all project that their industry will lead the charge into new interactive multimedia products and services. But multimedia is not a homogeneous market; indeed, it is not a market at all, but instead a process of adoption. The addition of multimedia capabilities fundamentally changes existing industries, creating new opportunities, threats, and confusion.

Changes brought about by new technologies require new knowledge and skills. Usually such skills and insights are acquired through direct experience and firsthand knowledge. Technological advancements, such as the transition to all-digital multimedia, produce a shift in existing business rules, causing a period of uncertainty and experimentation.

The sheer breadth of multimedia influence further complicates this change because the "multimedia" label is being applied to so many industries simultaneously—with entirely different meanings in each of them. In the case of multimedia, firsthand knowledge and experience in one industry probably is not enough to understand the new market rule; acquiring knowledge from and about other industries has therefore become critical to success and ultimately for survival.

The purpose of this book is to provide insight into some of the industries where, as with most families, there is disagreement among siblings, and the older generation tends to be less willing to accept radical change than the younger upstarts. Yet the stakeholders are being drawn closer together than ever before as the phenomenon of digital data creates, perhaps briefly, chinks in the armor of existing markets.

If digital multimedia provides potential openings among disparate industries, then intelligence about those other industries is of vital interest to business practitioners. This intelligence is not easy to come by. The obstacles that affect the new technology's acceptance and integration are frequently invisible to the scientists and engineers and strategic planners who design for the future. There is a basic understanding of the value and importance of multimedia technologies, yet there is a striking lack of familiarity with the business issues that provide a basis for sound decisions in any particular field.

So, the business of multimedia is about learning different businesses: understanding how they work, how they are different, and how they are the same. This book is designed for the architects of the new businesses that will emerge from the transformation to an all-digital information society. It is hoped that it will foster an understanding of both the opportunities and the obstacles embedded within each of several highly related but currently separate industries.

The Economic Structure of Multimedia

This book is premised on the notion that there are a number of separate but increasingly related multimedia industries, each with its own ecological systems. Multimedia information, consisting of any of video, audio, graphics, text, and animation, is common to these vertical business systems. The component parts of the digital multimedia information food chain are separately identifiable, but often bear little resemblance to their counterparts in other industrial "ecosystems."

Regardless of industry category, all multimedia content "flows" through similar process steps:

Capture → Development → Transmission → Display

The tools and services, transmission and display systems employed in various industries varies widely, but the processes can be identified in each industry. Differences in the display and transmissions systems among the various industries profoundly affect multimedia content design and capture. For example, a screen display designed for a computer usually will appear unacceptable on a television set. Interline jitter, color crawl, and a generally lower resolution can take a perfectly fine computer screen signal and make it unpleasant and unreadable. Similarly, video compressed for use on a desktop computer and delivered on a CD ROM usually will be fuzzy, lacking in color depth, jerky, and usually less than full screen—far, far less appealing than television quality video.

The economics of each industry govern the technological requirements for the various multimedia industries, each building legacy systems that cannot easily be changed or modified without significant cost and risk. Therefore, understanding the transmission-display systems for each of the industries can yield important insights to how and at what rate new technology will be introduced.

The Continuum

It is useful to visualize multimedia industries as a continuum of "ecostructures," arranged to reflect common attributes and elements. Entertainment-based businesses are grouped at one end of this continuum, productivity-based in the center, and communications-based on the right. This arrangement is not arbitrary; each vertical grouping is related to its neighbor in very important ways, as will be described shortly. The historical transmission-display systems in each of these groupings are fundamentally different, both in terms of cost and technology. These differences, as we shall see, lead to basic differences of philosophy among these groups, each of whom is jockeying to grab a leading market position.

Television

Broadcast television is among the most prolific purveyors of multimedia content. The conversion to all-digital multimedia streams is approaching, but has not yet reached television over-the-air transmission, and is just now beginning to occur within cable systems (mostly to squeeze more analog signals through the same system). Driving the conversion to digital formats are the impending advanced television rules from the FCC that will mandate that broadcasters convert to all-digital signals as a requirement to maintain their current allocated spectrum, new compression systems that allow more services to be delivered within the same bandwidth, and the possibility of new interactive data services that form the basis for additional revenue growth.

Despite the possibility of new information products and services that interactive television and cable systems might offer, these closely related industries base their revenues primarily on providing *entertainment*. Government regulations impose other service and performance requirements, but these industries make their money by bringing amusement, diversion, and distraction into living room television sets nationwide.

	Entertainment		Productivity		Communications	
Support Services	– – – – – – – – – – – – – – – – – – – LAWYERS – Pre/Post – Studios – Graphic Artists – Designers – Subject Exp. – Coders – Consultants – Visionaries					
Tools/ Authoring	– – –Avid – – –	– – –[Proprietary]– –	– – – – –Macromedia Director – – – – – – – – (Kaleida ScriptX) – – – – – –		Client/Server – –	[customer]
Content Develop.	Paramont PBS BBC	Electronic Arts Acclaim	Broderbund Putnum Lotus Andersen Dor. Kind. Voy – – – –Microsoft – – – –		[customer]	[customer]
Delivery/ Transmis.	Coax/Fiber R.F.	[Proprietary] Cartridge/CD/Floppy	– – – – – –CD-Windows – – – – – – – – – – –CD-MAC – – – – –		Ethernet TCP/IP Twisted Pair	ATM IDSN ADSL
"Player"/ Platform	S.A. Pioneer Philips GI	Sega Nintendo (Sony) 3DO	Dell Packard Bell CD-I Compaq Apple IBM SGI		Novell Banyon DEC SUN	BT Bell Atlantic AT&T U.S. West
	TV/Cable	Games	Home PC	Business PC	Enterprise-Wide Comp.	Telecom.

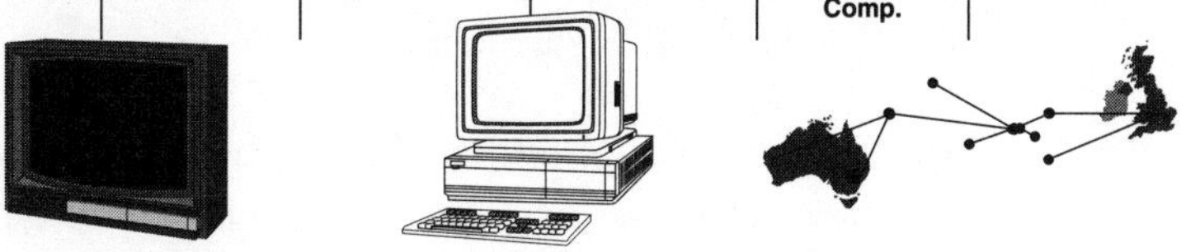

Figure 1.1   The multimedia continuum

Cable

The cable industry originally evolved to fill in the gaps between the range of broadcast television stations. These gaps occurred either because of the distance from broadcast antennas or due to geographic obstructions. Since early cable simply retransmitted existing broadcast signals, the new service seemed to be a natural part of the broadcast food chain.

Broadcasters did not see it that way. They long ago realized that cable was and is a threat to traditional television stations rather than an enhancement to their businesses. With cable, fewer stations and towers are needed to transmit high-quality programming into homes. Entirely new entertainment and information networks have developed in this environment, and, unlike over-the-air broadcast signals, which are highly regulated by the government, many cable systems were developed and deployed using entirely proprietary technologies.

Today's cable companies are well aware that the regulatory climate is closing in on them as it has for broadcast and telecommunications companies. Most are still reeling from the Cable Act of 1992, which the broadcast community lobbied hard for and won even over a presidential veto. Yet their independent and entrepreneurial approach to their business had sparked a new level of media hype, promising new broadband services and capabilities that can be delivered only by a wired infrastructure.

However, the promise of interactive multimedia capabilities and digital media streams is largely unfulfilled due to the lack of solid business models for new services. No one really knows what consumers are prepared to pay for. Leading candidates include video on demand, home shopping and banking, distance learning, and interactive multiplayer games.

All the major cable companies are rolling out tests to see if they can confirm consumer demand for these types of new services. The implications for each of these services have spawned dozens of industry technical working groups with the hope that standards may emerge that will lower the costs and risks of adding new services. Meanwhile, most cable companies continue to incrementally upgrade their wiring plant, add new channels, and continue their test marketing—waiting for the "killer application" of cable to emerge.

Video Games

Interactive game companies realized long ago that the television set is the primary vehicle for delivering entertainment to the home. By connecting directly to primary home display devices, they leveraged the installed base and simply provided consumers with the one feature missing in broadcast television: interactivity. Like television, video games provide only entertainment. Consumers' tastes, budgets, and habits profoundly affect the success of all entertainment-based products. Thus both television and video games are high flying, big hit markets with relatively short product life cycles.

Also in both cases, they are very big markets where fortunes can be made and lost in mere months.

Once a small part of the larger consumer electronics palette of products, interactive games have grown to the point where the game industry has now formed its own association separate from the mainstream consumer electronics market; it is called the Interactive Digital Software Association (IDSA). This organization is now dealing with legislative initiatives such as mandatory ratings for games and copyright protection—much like other parts of the entertainment industry.

Home Computing

For years computer products marketed to the home were connected to television sets just like video game machines. Early computers, such as the Apple II, provided connectors to television sets. Games figured prominently in early home computer software offerings and still dominate home software purchases. Devices used in the home and connected to the television set are expected to entertain. As a result, color, sound, and animation have always been a major part of the design criteria for successful interactive home products.

Home computing adds new levels of interactivity with a full-text keyboard, a pointing device, and storage media (e.g., diskettes), and thus promises to provide productivity gains. So, home computers begin to take on characteristics of both business class desktop computers as well as video game machines. As multimedia capabilities have been added to business computers, and prices have dropped, a new breed of home computers has now taken off, offering full compatibility with business machines and multimedia features that are comparable with game machines. Although the productivity capabilities of home computers provide a good purchase rationalization, entertainment (in the form of CD-based games) remains the primary consumer motivation for purchases.

These new home computers break from previous electronic entertainment tradition by providing their own display monitor intended for use on a desk top. These machines have generally the same architecture as computers sold to business, but have multimedia capabilities added in. To make these computers interesting to the home purchasers, manufacturers added high-capacity storage, in the form of CD-ROM disc drives, and add-in cards to produce sound, audio, and video. This new breed of home computer delivers entertainment, like television and video games, and offers the ability to bring work from home—a very compelling sales pitch.

The commonality of home computers with their business desktop parents created the opportunity for a new category of products: home education. Because home computing straddles entertainment and (what most parents hope their children are productively engaged in) educational activities, some call the new software category "edutainment." A whole new series of fun software products designed to be healthy alternatives to passive television and mindless video game products have hit the market and been snapped up by responsible parents who want their kids to have the best competitive chance in school. Replacing bulky book sets, CD-ROM reference titles include dictionaries and full multimedia encyclopedias. Besides, when the kids go to bed there are some really nifty games that adults can play too.

Business Desktop Computing

Over the past decade, the business desktop has been the launching pad for entirely new hardware and software empires. Leveraging each generation of more powerful microcomputers, new productivity software for creating documents, managing finances, and managing data exploded onto the market during the early eighties to empower individual users who formerly had to rely on centralized service organizations to get their work done.

The desire for individual productivity and tools created a strong vacuum for new products, many of which were, in their early incarnations, expensive, difficult to use, and sometimes unreliable. Nonetheless, advantages outweighed disadvantages and the market took root quickly and grew. Rapid revisions to productivity tools and platforms created a kind of addiction among business users for the "latest and greatest" product releases, and established new criteria for adequacy and professionalism through the addition of new capabilities and features. For example, soon after relatively low-priced but high-quality laser printers hit the market, lower quality dot matrix printouts became an indicator of an amateur or nonprofessional organization. Business computing, in both hardware and software, rapidly became a game of one-upsmanship, continuing to climb on the back of real or imagined productivity gains.

The design objectives and primary use for business computers have always been crystal-clear: they are tools for the creation, management, and manipulation of business information and data. They are complex, creative environments that arrive empty, ready for users to pour in their personal data. Then, the information is "baked" and "served" on high-quality graphics screens or printers.

Business users have tended to resist multimedia capabilities from the early days of personal computing. The transition from monochrome to color monitors and the conversion to graphical user interfaces took longer than expected because many corporate buyers felt that they didn't need color or graphics for their business. Gradually, as prices fell, color and graphics capabilities came built-in at no extra cost. No longer optional, these capabilities ushered in new, innovative tools and features. Now, multimedia capabilities are being gradually added to the recipe, with the same initial corporate reaction: "I don't need video or audio in my business." But multimedia-capable desktop computers are rapidly becoming the norm in businesses and it is ever easier to imagine new applications and uses for these products as the installed base of such systems grows.

Already building on the strong legacy as a platform for productivity tools, multimedia-capable desktop computers have moved into the supply side of entertainment-based industries, including film, television, and video game development. Movie producers are now converting wholesale to digital special effects created entirely on leading edge work-stations, and saving millions in film production dollars. Desktop computer editing systems for television, film, and audio now exist at a fraction of the cost for dedicated post-production equipment. Despite its use as a creative tool in the entertainment industry, the role of the business desktop computer remains principally as a manipulator of externally supplied multimedia data rather than the delivery mechanism for prepared multimedia information, objects, or experiences.

Home computers, which usually lack the raw horsepower (and therefore the costs) of business desktop computers are shifting to the "player" role, and are increasingly used to render multimedia content that is not designed to be edited, altered, or made part of a productivity project. Rather, such content is often designed to be digested "whole" although with some measure of interactivity to navigate through linear "content" segments. Thus multimedia divides the computer bloodline into productivity, tool-based platforms designed to manipulate information that is put into the system, and multimedia player systems adept at searching and rendering multiple datatypes in predesigned, synchronized object events. In other words, tool devices and player devices.

Enterprise-Wide Computing

The term "enterprise-wide computing" is meant to convey an interrelated set of information and computing services implemented throughout a large organization. If organizations are likened to biological organisms, then the computing system is the nervous system and major parts of the brain (memory, computing, communications, etc.). The systemic aspects of this highly evolved environment dominate the design and philosophy of products and services in this category.

Early mainframe-based centralized computing centers inevitably needed to offer remote connections to provide the ability to input or output data from geographically distributed business centers. In order to serve as the organization's nervous system, information had to flow to and from major organs and limbs. For this reason, management information infrastructures have always had a strong communications and networking component. As more powerful and smaller products were introduced, the web of connections increased and many workstation products were and still are delivered with all the hardware and software necessary to connect to corporate networks. In contrast, personal computers have operated in relative isolation, requiring add-in components and software to achieve limited connectivity.

The emphasis and focus on enterprise interconnectivity established a significantly different philosophical framework than stand-alone computing. For one thing, the design of internal computer architectures and capabilities is secondary to its ability to integrate into the overall business system. The various nodes of a complex system each provide different capabilities and services, but the sense of their individual value is diminished in comparison to the overall network itself. A kind of "connectivity culture" has now developed, relying on and fostering open architectures and interoperability throughout the system. This culture is diametrical to the stand-alone computer community, whose idea of infrastructure is defined fully within their own computer boxes.

Along the distant desktop borders of the enterprise computing empire, alliances formed on the low-cost end of the business. As the horsepower of stand-alone computers accelerated, such systems became more attractive to users than embedded and centralized systems. The lower-cost products soon were able to perform many of the same tasks, but now at a cost that was attractive to a huge new market of small and medium size organizations. All these computers lacked was connectivity. Enter the local area network (LAN).

Low-cost local networks vaulted small groups of computer users into a new electronic community that began to challenge the need for more expensive centralized approaches in many cases. This community developed its own standard and imposed new rules. Once connected by a network, individual systems became hobbled and sometimes lost the differentiation of features and performance as they all shared the same printer with the same throughput, and were forced to use the same networked applications in order to exchange information and maintain software compatibility.

More recently, local networks are being connected to one another and to wide area networks within distributed organizations, all of which is part of the current "downsizing" and redistribution of corporate information systems. Still the emphasis is on data interchange and mobility rather than individual computing capabilities. Connecting existing desktop computers of different types with different architectures and operating systems is now commonplace along with software applications with multiple native clients for use on a variety of systems from a common server.

One effect of enterprise-wide computing has been the development of electronically defined communities. This is spawning new work patterns as work group software aggregates information along topical rather than geographic lines. The computing portion of information systems, that is, the raw database crunching of past information systems, is now being overtaken by communications processes designed to connect people to one another.

Enterprise-wide computing includes wired microcomputers, workstations, mainframes, servers, coaxial wire, fiber optic cables, routers, gateways, interfaces, protocols, and hundreds of other bits of electronic "glue." Network capacity and bandwidth both enable fabulous new capabilities while imposing a hard limit on system-wide throughput. The complexity and costs of such systems mean that they are not easily or cheaply modified and altered to add new capabilities.

So, the role of enterprise-wide computing is evolving toward information exchange between people (as opposed to data storage and retrieval). Multimedia plays an important part in this new role since people need to exchange real world information and communicate best using the primary human senses (sight and sound). The capabilities or limitations of individual computing platforms are less of an issue in the adoption of multimedia capabilities in connected environments than is the nature and capacity of the network that binds them together. Multimedia data's huge bandwidth and synchronization requirements far exceed what is available from today's local or wide area networks. Thus the networking and telecommunications issues outweigh computational challenges as multimedia capabilities are brought into the enterprise information systems.

Telecommunications

It used to be that the words *multimedia* and *network* were never uttered in the same sentence. The bandwidth requirements of audio, images, and video have always posed a threat to the capacity of nearly every type of wired data transport in existence. Even broadcast television, the basis of very different kinds of networks, was long ago declared a "scarce resource" that needed to be managed and watched over closely by governments.

Now, due to increased speeds, new compression technology, and dramatic decreases in data transport technologies, networked multimedia is a common topic. The issues are still huge, however, and the business models are still evolving. The factors that govern the evolution of networked multimedia now are being driven by the evolution of new business models as well as changes in the regulatory environment — not, surprisingly, by technology advancements.

Defined as the technology of transmitting information in an electromagnetic form, the earliest forms of telecommunications began as coded signals and now include voice (radio and telephone) and images (facsimile and television). More recently, data communications make up a growing proportion of electronic communications. In a sense, telecommunications long ago shifted from "data" representation to multimedia-rich formats, and now seems to be moving back toward coded (digital) data. Some of the reasons for these shifts are fairly obvious: older multimedia-rich telecommunications systems tend to be "point to multipoint," or broadcast, whereas interactive services are necessarily "point to point" communications. The economic differences between a one-way "source" system with passive receivers and full duplex communications are vast and help shape the fundamental difference between broadcast programming and what we think of today as our communications infrastructure.

Telecommunications systems primarily transport information in real time. They have been historically designed to put people in contact with one another over great distances. The business is the bridging of physical distances with an electronic "pipeline." Telecommunications providers have either been relatively uninterested in or legally precluded from involvement with the traffic that flows through their pipelines, and have instead emphasized system reliability and multipoint access over developing new types of services and capabilities.

The "diameter" of the pipes for most of the global communications system has been set for the human voice, or actually less than the human voice's spectrum in terms of bandwidth, but still enough for people to be clearly heard and understood. With a limited bandwidth pipe, the cost per mile for laying these pipes is lower than if the pipes had to carry more information (e.g., copper wire pairs and their associated amplifiers and switchers are cheaper than coaxial cable systems).

The conversion of telephone systems from analog to digital audio signals produced manifold improvements and cost efficiencies, but did not affect, in any material way, the basic service that could be used or seen by the average customer. Digitally multiplexing existing pipelines vastly increased the traffic over existing lines and significantly improved signal quality. Still, basic telephone service provides a real-time, near voice-quality bandwidth connection, and little more.

Though designed for voice, facsimile machines and computer modems utilize the analog telephone transmission medium for digital information. To accomplish these communications feats, digital bits are translated into "analog friendly" signals, sent over audio lines as frenetic screams and shrieks, and then reconstituted into usable digital data again. The limited bandwidth available over voice lines, together with the conversion and error checking overhead of converting digital to analog and back to digital again, creates a kind of "speed of light" barrier where with increasingly sophisticated technology allows telecommunications-based products to add more and more multimedia capabilities, without ever being able to reach full multimedia capabilities over existing telephone wires.

Demands for high-bandwidth data communications have, of course, triggered the development of new data pipelines that can have comparatively limitless capacities. These systems form the backbone of both voice and data communications systems. However, these high-volume digital pathways are not cheap, and are out of reach of the typical individual consumer or smaller businesses. ISDN, ADSL, or ATM service, which deliver orders of magnitude more data capacity than "plain old telephone" service, carry costs many times more than standard voice service.

At least until very recently, services provided by telecommunications companies have been computationally silent. That is, the service provider neither acts upon data transmitted (except during the initial connection and billing processes) and there is no expectation of a computed process or protocol to maintain basic connection. Sophisticated digital communications systems usually establish data connections and issue complex communications protocols only after the basic telecommunications connection has been made. This means that telecommunications services have been relatively dumb and have not been expected to send or receive complex data.

Multimedia Networks

Business and consumer multimedia network markets have markedly different economic and technical criteria for success. Even though networking issues seem blurred and could be treated as abstract problems that apply to both business and consumer applications, the reality is that these are two very different environments and markets.

Business networking provides private information and data transport. Each component of an enterprise system is built to solve a particular corporate need and must be justified on the basis of the contribution to the overall business.

Therefore, the addition of multimedia capabilities must be mapped to real business needs that are cost-justifiable and that are fully compatible with existing information-technology infrastructures. Some examples include computer-supported collaboration, which is basically an enhanced communications medium, and photo-image networks for news magazines or other publications whose business depends on quick access to high-quality multimedia data.

There are, of course, hundreds of viable business applications that are fundamentally multimedia in nature. However, the differences among the businesses give rise to entirely different applications requirements. The result is that businesses are *evolving* incrementally, each at its own rate according to its own needs. Step by step, existing applications are being adapted to add multimedia capabilities in business applications all over the world, but they are not the *same* capabilities, nor do they necessarily employ the same technologies.

The multimedia value added is coming mostly from the enterprise-wide community, as defined by the continuum model. Telecommunications companies, for the most part, have been interested in (or limited by regulation to) data transport only. The notion of multimedia has, to them, been a vague concept—a nice idea that ought to drive demand for bandwidth in the future. Enterprise-wide, business-oriented organizations operate on an intimate level with their information infrastructures, which are in turn welded closely to individual corporate structures. It isn't surprising that these organizations, rather than the telecommunications companies, are leading the way in adding business-related multimedia capabilities.

The adaptation of existing business networks to multimedia has been difficult. Only during the past few years have the detailed requirements for distributed multimedia applications begun to be fully understood. To address some of these issues, the Interactive Multimedia Association (IMA) initiated work in its Compatibility Project and developed a Request for Technology for Multimedia System Services for distributed environments. The resulting submission and model illustrates the challenges and changes that multimedia demands of today's system.

Consumer Multimedia Networks

Threatened competition between telecommunications and cable companies, combined with decreases in technological costs and barriers as well as the probability of relaxed regulation, is driving the construction of vast new consumer multimedia networks. The products and services offered by these groups are certain to be aimed initially at the general public and then at vertical niche markets. Cable and telecommunications companies cannot afford to lose their primary customer base.

The classes of services that are likely to drive this sector are, of course, entertainment and consumer services (e.g., shopping, banking, etc.). Even though the successful service models are not yet fully understood, the broad categories of capabilities that consumers value are becoming clear, as well as what they will use and pay for.

What is not understood by many is that the deployment of sophisticated interactive multimedia networks for consumer applications *is not gated by technology any more.* Instead, service models, which define a basis for investment together with new government regulations, are having the largest effect on technology and topology decisions. The classification system, which is described below, applies principally to consumer-targeted network systems (currently dominated by cable companies, but under intense scrutiny by telecommunications companies). It illustrates the complexities of service-based infrastructures.

Multimedia Network Classification

The IMA Set-Top System Classification was created as a simple way to describe various set-top related infrastructures that appear to be relevant to a variety of potential interactive media service providers. This taxonomy focuses on the overall *system* rather than set-top devices or specific technologies. It is designed to describe the basic environments in which various services might be deployed. It is understood that each of the classes will need to be subclassed before technical capabilities or requirements can be fully defined.

The dividing lines for each of the classes take into account existing networks that may become the basis for new services as well as new topologies that are in planning stages. The definitions are also intended to be a useful categorization of network systems when attempting to distinguish among various service types currently or prospectively under government regulation.

Medium and Back Channel

The classification of Set-Top systems is organized principally around two parameters: the medium in which data and content are represented (analog or digital), and the degree of availability of a return channel of data to the head-end or server. More than other characteristics, the nature of these two elements governs the capabilities of the overall system.

Class 1: Analog video channels with no return channel. This covers most existing cable television systems. Channels are at predetermined frequencies.

Class 2: Analog video channels with a minimal and typically very slow return channel. The back channel might be used for simple set-top device identification and/or simple pay-per-view requests. Interactive services are not supported; back channel technology is often proprietary and part of a monolithic end-to-end network system.

Class 3: This hybrid class is also built on an analog-based wiring plant, splitting the spectrum to include mainly downstream digital information such as encoded video like MPEG, and/or allocating existing channel bandwidth to encoded digital information. The back channel in this class is minimal, as described in Class 2. A digital signal decoder is presumed to exist on the customer site for the digital streams.

Class 4: Also hybrid (digital overlaid on an analog network), but with modifications to the wiring plant to allow a reasonable data return path. This is the first really interactive system. The return path is of much, much lower bandwidth

Table 1.1   IMA Set-top Systems Classification

Class	Medium	Return Channel
Class 1	analog	none
Class 2	analog	minimal
Class 3	hybrid	minimal
Class 4	hybrid	asymmetrical
Class 5	digital	asymmetrical
Class 6	digital	symmetrical

than the downstream data and is therefore labeled "asymmetric." As with all lower classes, this class is presumed to support standard clear-signal analog television channels for connection to so-called cable-ready television sets. Class 4 splits the spectrum cleanly between analog and digital, offering additional and perhaps incremental digital interactive services over and above the standard analog service. Considerable variation in the return channel bandwidth as well as latency is expected in this class, thus giving rise to the need to subclass for specific service types.

Class 5: All digital. Like Class 4, the data path is asymmetric (more downstream "broadcast" media than upstream data), although an all-digital system hints at the possibility of higher bandwidths enabled by high performance switching and transport mechanisms. This class requires a decoder device for each television set to translate digital signals into cable-ready channels or computer screen-ready display.

Class 6: Full duplex. This is an all digital network with the same bandwidth upstream as down. At the lowest speeds, this is comparable to today's data networks (e.g., Ethernet). At high bandwidths (in the distant future) full video telephony is supported. Classes 5 and 6 have no analog signal transmission anywhere in the network and therefore differ from the other classes at a transport level.

Implications

Business applications for networked multimedia will develop rapidly as relatively isolated islands. These islands will have different capabilities and degrees of multimedia content and will mirror the organizations they serve. These systems are not likely to be particularly interoperable in the near term, although that is probably unimportant for the most part.

The major growth of multimedia networks will come from telecommunications and cable companies seeking to protect and grow their consumer products and services. Entertainment and new interactive services will drive the topologies along with government regulatory changes. Over time, as the scale of the investments increases, the consumer networks and technologies will back-feed vertical applications in business.

Finally, through the combination of consumer and business network evolution, relatively high capacity will become available on a widespread basis by the end of the decade, thus driving the need for symmetric, synchronized networks with reliable bidirectional bandwidth.

Impact on Multiple Industries

All the vertical industries in the multimedia continuum are harnessed with their own economic restraints and histories that operate to limit the ways in which they can take advantage of new media trends. As everyone struggles to understand these new markets, they face new sets of issues and new roles that are culturally jarring and often dislocating.

The entertainment industries face the edge of an entirely new digital media infrastructure that will directly impact the entire food chain of what they call "software" from creation through to all-digital high-definition displays in homes. In addition, they face totally new transport systems, distribution media, and, least understood of all, interactivity.

Providers of productivity systems are contending with the shift from tool provider to consumer appliance maker. The cost structures, margins, inventory issues, marketing requirements, and sheer volatility of consumer markets are completely alien to many traditional business computer makers, any number of which will no doubt fail in an attempt to serve the new multimedia markets. For their part, a number of workstation companies have shifted a substantial part of their business to serve the creative community—the so-called top of the food chain.

Network providers face huge opportunity and risk, moderated by legislative limits that may well be removed or significantly altered in the next several years. Cable companies will be faced with defending their turf from telephone companies entering their business, just as cable adds communications capabilities. Meanwhile, both groups grapple with the implications and the viability of interactive services that mandate huge infrastructure changes.

The Never-Ending Story

There is, of course, no possibility of creating a definitive description of the evolution of the "multimedia market." Nonetheless, extrapolations are possible when a number of statistical points are taken. Each of the following chapters is designed to be a glimpse into one very narrow aspect of multimedia and to provide data points on which one may build intelligent guesses about the ways in which multimedia are likely to develop in a variety of businesses.

Contributing authors were asked to address specific topics in a specific order and to educate the reader about their own little corner of the business. Some attacked this process rigorously; others struck out in their own directions. We hope the diversity will serve to enlighten rather than confuse. The subjects posed are as follows:

1 Industrial history and memory

2 Profile of the user

3 Distribution systems (reaching the user)

4 Core technologies

5 Applicable standards

6 Infrastructure (current/planned)

7 Regulation and legal constraints

8 Industry outlook

These topics hold great insight into the inner workings of each of the industries and how they are likely to evolve. The mind-sets of the players as much as the actual environmental conditions profoundly affect the business decisions that will shape the new media digital world.

2

Broadcast Television

MICHAEL C. RAU

A cross-industry multimedia tour rightfully begins with broadcast television, the oldest and most experienced of multimedia-based electronic infrastructures. Many of our perceptions and prejudices as consumers emanate from our nearly lifelong experiences with the "one-eyed monster" at home. This chapter traces the earliest demonstrations in 1939 through to the next-generation conversion to digital broadcast technology, high-definition television (HDTV).

The role and degree of involvement of the government (especially the FCC) is particularly relevant to architects of new media systems. Throughout its history, the broadcast television industry has been yoked to government for setting standards, allocating spectrum, licensing, and now, converting to all-digital signals. Unlike other industries that will be profiled later, broadcast TV seems relatively comfortable coexisting with a government that sets the rules for all. The mandate to convert to HDTV as a requirement to continue broadcasting over the next 15 years would appear to be an unusual degree of government meddling in free-market affairs. Nonetheless, the history of broadcast suggests that such intervention can be a highly effective way to both protect consumer interests and grow a huge new industry.

In terms of sheer size, few markets are as large as that which broadcast claims to have penetrated: the entire population of the United States. The business model that fuels this industry, however, comes principally from a group other than consumers—advertisers. Using an old and implicit contract with consumers, broadcast content is created and delivered free of charge, in exchange for the insertion of advertisements. Relatively new cable services, such as the Discovery Channel and the Arts & Entertainment Channel, are proving that advertising is very much a viable business model even when there is also a basic charge for the transmission/cable service. Who pays for the content (advertisers, consumers, or a bit of both) establishes the fundamental business model differences for the various players in multimedia.

Though many view broadcast as an industry in decline, the greater probability is that it will be the broadcasters who will drive convergence of all the other industries through an orderly adoption of externally mandated standards on a fixed timetable. This changeover to all-digital media may prove to be the primary driver for the creation of low-cost digital appliances that will replace the old analog television set we grew up with (history certainly bears out such a prediction). With more than 60 years of experience, we may yet have much to learn from those old broadcasters after all.

History

Definition

Webster's dictionary defines broadcasting as "to make known over a wide area" or "to scatter over a wide area." Over-the-air broadcasting, then, is the dissemination of video and audio "over a wide area" Certainly, the *business* of over-the-air broadcasting, as we know it today, involves the acquisition or creation of programming, the delivery of that programming free to viewers, and the selling of advertising to pay for the costs of programming and station operations and, we can hope, provide a reasonable return on investment.

The Birth of Broadcasting

Television broadcasting, as you might expect, has its origins in radio. More specifically, the concept of broadcasting or electronic mass communication began in the form of radio broadcasting. Guglielmo Marconi is generally thought of as the father of modern wireless communications. He successfully sent the first radio message in 1895 and then in 1899 sent the first long-distance transmission, in Morse code, across the English channel. But, Marconi's invention was really intended for telegraph use, not for the general public, and as such really does not fit into the definition of broadcasting. Marconi's work, however, was a significant step toward radio and eventually television.

Perhaps one of the most important technological advancements that led to the development of broadcasting was the invention of the Alexanderson Alternator. This device, manufactured by General Electric, was the first high-power, continuous-wave transmitter. It was built at the request of Reginald A. Fensenden who, using the Alexanderson Alternator, made the first radio broadcast on Christmas Eve 1906 from Brant Rock, Mass. Fessenden's transmissions consisted of voice and music, had a regular schedule of programs, and was intended for the general public, at least anyone with a receiver.

The Alexanderson Alternator is also significant because it was the center of a dispute between GE and British Marconi. It was the ensuing negotiations that led directly to the formation of the Radio Corporation of America, RCA. As a result of those negotiations, GE purchased a controlling interest in American Marconi. Having no interest in running the radio facilities owned by American Marconi, which consisted mostly of ship-to-shore maritime stations, GE created RCA in

1919, so that it could concentrate on manufacturing radio equipment. Two of the officers of American Marconi, Edward J. Nally and David Sarnoff, became, respectively, president and commercial manager of RCA.

Shortly after the signing of a number of patent agreements among GE, RCA, Westinghouse, and AT&T, radio broadcasting, as we know it, was born. A final element worth noting was the airing of the first advertisement. The first commercial was aired on WEAF, owned at that time by AT&T, in New York City on August 28, 1922. The spot was purchased by the Queensboro Realty Company for $100 and was 10 minutes long. At that time, AT&T referred to the selling of ads as Toll Broadcasting, as a parallel to toll telephone service.

Television's Beginnings

Fast forward about 10 years. David Sarnoff was now at the helm of RCA; AT&T had, for the most part, removed itself from the radio broadcasting business; and America was coming into the golden age of radio. After years of development and millions of dollars invested in research, RCA mounted the first public demonstration of television in America on April 30, 1939 at the New York World's Fair. It was clear that television was nearing the point of commercialization. However, the FCC indicated that they would not take action on a television standard until there was general industry consensus on what that standard should be.

In July of 1940, the Radio Manufacturers Association established the National Television Systems Committee (NTSC) to create a transmission standard for monochrome television. The NTSC was chaired by Dr. W.R.G. Baker. The NTSC delivered its final report to the FCC on March 20, 1941, recommending a television standard. The FCC officially adopted this standard on April 30, 1941 and ruled that commercial broadcasting based on the NTSC standard would be permitted after July 1, 1941.

Fast forward again to 1950. The FCC had adopted a color television system that was not compatible with the existing monochrome system, against the advice of a great number

of the industry's technical experts. As a result of this contro-
versy, the NTSC was reactivated in 1950 in order to develop
a consensus on a color TV system. By July, 1953, the NTSC
had developed that system. It was adopted by the Committee on July 15, 1953 and submitted to the FCC the next day.
The FCC approved the color standard on December 17,
1953, authorizing color transmissions to begin after January
22, 1954.

Television Today

There are now over 1,500 television stations on the air in the
United States. Approximately 75% are commercial stations,
with the remainder being noncommercial, educational stations. Much of the recent growth of television stations has
been in the UHF band. Growth of stations between 1980 and
1991 is shown in Figure 2.1. Broadcast employment between
1980 and 1990 is shown in Figure 2.2.

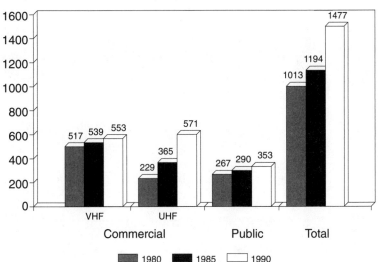

Figure 2.1 **Growth of U.S. television stations between 1980 and
1991.** *Source:* **NAB Broadcasting Profile.**

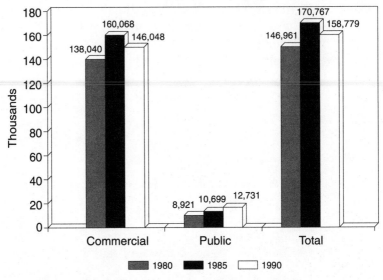

Figure 2.2   Full-time broadcast employment between 1980 and 1990. *Source:* **NAB Broadcasting Profile.**

Impact of HDTV and Digital Broadcasting

In the increasingly competitive communications market-place, broadcasters must offer their audiences a combination of the best technology and programming. Adherence to this goal has led to the current arrival at the final stages of setting a digital terrestrial HDTV broadcasting standard for North America.

HDTV has four basic features which distinguish it from present formats:

1 Twice as much horizontal and vertical resolution as NTSC;

2 Greatly improved color rendition;

3 A 16:9 aspect ratio wide screen display; and

4 Multichannel digital sound.

The new digital world associated with HDTV broadcasting offers opportunities far beyond just better pictures and sound. Flexibility to implement other digital services along with HDTV will open many opportunities for broadcasters. These services may be available either while transmitting HDTV signals or at times when the HDTV signal is not in operation. The ability to transmit ancillary data to a new generation of digital receivers—from computers to fax machines to personal digital assistants (PDAs), personal intelligent communicators (PICs), and other receivers—can potentially help broadcasters generate additional revenues to pay for the transition to HDTV. Due to the flexibility potentially offered by digital broadcasting, the general term, advanced television (ATV) is often used instead of the specific service application, HDTV.

Flexibility also gives broadcasters the opportunity to participate in the multimedia world of tomorrow. All this is possible because the advanced television process allows broadcasters to make the transition from the current analog service to a digital service. The FCC recognized that consumers cannot be forced to switch immediately from analog to digital in their approach to HDTV. Rather, for a potentially lengthy transitional period, broadcasters will be required to provide the same service on two channels (one in analog NTSC and one in digital HDTV). The rationale for this simulcast approach is that eventually the consumer will migrate from the NTSC program to the higher quality HDTV program. The FCC intends to eventually reclaim the old NTSC channel after the digital service is ubiquitous. Digital broadcasting using a second television channel is in fact the *only* way that broadcasters can migrate to digital services. If the NTSC service were suddenly terminated in favor of a new digital service on the existing channels, the entire population of 200 million television receivers in viewers' homes would be made instantly obsolete!

Profile of the User

The difficulty in profiling the television audience in the United States is sheer size. By any measure, TV is the most pervasive medium in the country. In the foreword to the

Markle Foundation's historic study, *The Public's Use of Television: Who Watches and Why*, Lloyd Morrissett, president of the John and Mary R. Markle Foundation, says: "Television audiences are so large that it is hard, almost impossible, to think about them." That study then attempts to catalogue the various reasons that people use television. The result is a complex, multidimensional matrix of user classifications that indicate television serves a myriad of uses from maintaining awareness of current affairs to obtaining validation of life choices by observing role models in TV programming. Most viewers reported that their rewards included a number of the different choices so that a "normal" viewer profile cuts across a number of classifications. The final conclusion of the report is that viewer types can be identified as a series of "clusters" that form a matrix around certain key shared interests.

The Markle matrix is so complex simply because the use of television is nearly universal. In the spring of 1993 Nielsen Media surveyors found that more than 98% of American homes have a TV set. In fact, 64% of households have two or more sets in active service. By contrast, only about 94% of American homes have a telephone. This means that the "TV Universe" in the United States includes more than 196 million people. The number of TV households grew steadily from 85 percent in 1980 to the current figure of 98 percent, which has remained stable since 1989.

Television's audience is, quite simply, the entire population of the United States. It is difficult to find an individual who does not have ready and immediate access to a TV receiver. The question then becomes, are they really watching the sets they own? Based on the latest ratings, between 8:00 and 9:00 P.M. on an average night there will be at least one television set in use in more than 60 percent of American homes. These same ratings indicate that the average American household uses TV for an average of about 8 1/2 hours every day. These figures have remained almost constant since the 1970s.

The primary change in television since 1970 has been the menu of choices available to viewers. At the beginning of that decade, many cable systems were still in the building

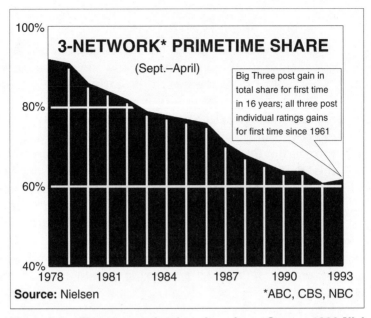

Figure 2.3   Three-network prime-time share. *Source:* **1993 Nielsen Television Survey.**

mode. Thirty years later we have reached a point where FCC Chairman Reed Hundt contends that 90% of American households have access to cable, and most viewers have adjusted to the increased viewing choices. This is clearly reflected in Figure 2.3. As cable systems were built in the early seventies, viewers began to take advantage of their new choices, and the audience share for traditional over-the-air programming dropped each year. By the last quarter of the eighties, cable build-out was finished and the share change stabilized. In the November of 1993 sweeps, this trend actually showed a minor reversal as broadcast TV programmers aggressively counterprogrammed against cable.

As the largest presence in American life, it is logical that television is the primary conduit for information into the American home. This is especially true for advertising. TV has been a critical part of product marketing since the early sixties, and that continues to be the case. According to recent reports, TV advertisers spent in 1993 a record $30,584 million. This came very close to equaling the $32,025 million spent in newspapers, easily exceeded the $27,266 million spent for direct mail, and dwarfed the $1,090 million spent on outdoor advertising.

The influence of television advertising is also showing a resurgence in the nineties. Despite TV's dominance, for almost a decade beginning in the early eighties, there was a slow but steady migration of advertising dollars from television to other media such as direct mail, couponing, and direct consumer promotions. This trend reversed itself beginning in the first two quarters of 1994. Most analysts credited this reversal to a steady drop in consumer spending that closely tracked advertisers' diversion of money from TV. Attempts to counter the consumer trend triggered growth of 9 to 13% in combined local and national sales income for TV stations in the first two quarters of 1994.

The primary arbiter in directing the flow of television advertising spending are the national audience ratings. Until 1993 there were two dominant rating services, the Arbitron Company and A. C. Nielsen, Inc. In 1993, Arbitron abruptly abandoned its TV service, leaving Nielsen as the dominant supplier.

The Nielsen company began business in the 1920s as a testing laboratory for industrial machinery. Acting on a tip from a pharmaceutical company in the 1930s, they expanded their business to include statistical studies of store stocks as a guide to consumer purchasing. As radio advertising became a major influence on consumer demand, Nielsen sought to develop automated methodologies for measuring listening patterns. The first home "Audiometers" went into service in 1938 and were quickly adapted for television as the new medium began entering American homes. A more sophisticated version of this same technology is still used today.

A full explanation of the enormous variety of reports and profiles produced by the ratings service is the subject of numerous books and the basis of an entire industry of consultants. The crucial question here is, why are these ratings so important? Quite simply, ratings are the only equity that the television industry can offer to advertisers. They are literally the "gold standard" for the industry. This concept needs some explanation. Begin by considering what advertisers are buying when they buy a television ad. They hope to buy effective promotion for a product or service. Unfortunately, it is difficult to accurately gauge the effectiveness of advertising with a given broadcaster after the advertising has run and virtually impossible to predict the effectiveness in advance. Someone once said that broadcasters are selling "Blue Sky and promises." Given these uncertainties, advertisers demand some basis for their purchase decisions. In the absence of anything firmer, the best indicator the advertising and broadcasting communities can use are audience measurements. While these "ratings" provide no guarantee of predictable advertising results, they are at least an indicator of who will be exposed to the advertiser's message and provide some measure of justification for the purchase decision.

There have been countless arguments offered and reams of paper printed in a continuing debate about the impact of ratings on television programming. One-time Federal Communications Commission Chairman Newton Minnow justified his agenda for funding the Public Television System by contending that the economic pressure to produce programming that would garner the highest ratings had produced a selection of look-alike programs that were little more than a "vast wasteland." In spite of the intellectual furor triggered by Minnow's criticisms and the introduction of alternative programming from public television, cable, and home videotape, the average American household still spends a little over eight hours every day with television.

One factor that continues to fuel the public's use of television is its ability to provide immediate news and information service from virtually anywhere. This ability continues to grow as TV equipment becomes smaller, and the worldwide network of satellite and other support technologies continues to expand. The impact of this ability is the key to understanding the public use of television. Television is the most pervasive information source in American life. There have been numerous studies that have attempted to determine the exact effect this "instant information" has produced. The results are indeterminate at best. Scholars lament the fact that the time restraints of TV news reports limit the depth of information that can be presented, and that TV's dependence on image can make it subject to manipulation by those who are skilled in "packaging" a person or idea in a positive way. At the same time there is abundant praise for the power of a medium that could end a questionable war effort by bringing graphic pictures of the conflict into homes only a few hours after the events took place. It is difficult to make a truly fair assessment of television's social impact. Long-time TV broadcaster Dave Garroway observed that "Television is like a knife. In the hands of a skilled surgeon, the knife is a scalpel that saves lives while in the hand of a felon, the knife is a weapon that kills and destroys." The impact of television on politics and other aspects of American life depends strictly on the skill of those who use it.

Program Distribution Systems

A completed television program is usually in the form of a videotape. The program may have been produced by a Hollywood studio, an independent producer, or directly in a network studio. The tape is delivered to the broadcast network where the network's national advertising is added. If the program is intended for syndication, the tape will be sent to the syndicator or arrangements are made to have the program fed via satellite directly by the producer. Syndicators will insert national or other advertising. Syndicators may also offer the program for barter. A barter, as its name implies, involves a trade: the syndicator trades (rather then sells) a program for a specified amount of local advertising time in that program. The syndicator will then sell that ad time to other national advertisers, agencies, or other interested parties. Barter ads may be inserted anywhere in the syndicated program distribution chain.

Local broadcast stations receive their programming from the networks and syndicators, or may purchase programming directly from other sources. The local station will also produce its own live and taped programs such as local news or sports as well as public affairs. Stations then mix in local advertising, usually as the program is being broadcast, for delivery to viewers. An overall view of this program distribution chain is shown in Figure 2.4.

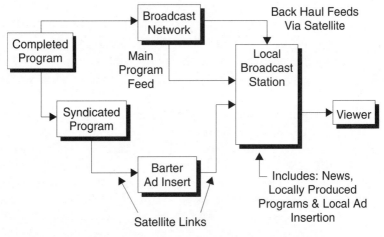

Figure 2.4 **Television program distribution.** *Source: NAB Engineering Handbook,* **8th Edition.**

Core Technologies

NTSC

Basically, television is a system of communications consisting of the television station at one end of the system and the television receiver at the other. Very simply, the function of the television station is to divide and subdivide the optical image into over 200,000 discernible picture elements, each of different light intensity; convert these light elements to electrical equivalents; and transmit them in orderly sequence over a radio-frequency carrier to the television receiver. Reversing the process at the receiver, these electrical signals are each converted to light of corresponding brightness and reassembled to produce the transmitted image on the face of the picture tube.

Picture elements to be transmitted in sequence are selected by a process of image scanning, which takes place in the television camera focused on the studio scene at the station. Within the camera, an electron beam in a pickup tube scans a sensitive surface containing an "electrical image" of the scene of action. The electron beam successively scans the image at great velocity, beginning at the upper left corner and continuing left to right in a series of parallel lines to scan the image completely. Movement of the electron beam, which can be controlled magnetically by vertical- and horizontal-deflection coils surrounding the tube, is analogous to that of the eye in reading a printed page. The speed of movement is such, however, that 30 complete image frames of approximately 500 lines each are scanned every second. Of course, at the receiver, an electron beam in the picture tube moves with the same speed and in synchronism with the camera-tube beam so that the corresponding picture elements appear in the proper relative position on the television screen. Owing to persistence of vision and the speed of scanning, these elements appear to be seen all at once as a complete image rather than individually. Thus, the impression is one of continuous illumination of the screen and direct vision.

Scanning standards have been established in this country to assure that all television receivers are capable of receiving programs broadcast by any television station within range. The scanning pattern adhered to by manufacturers in the design of television receivers and broadcast equipment consists of 525 lines with odd-line interlaced scanning. Interlaced scanning, effective in eliminating perceptible flicker, is a method whereby the electron beam scans alternate rather than successive lines. For example, the beam begins by scanning odd-numbered lines (1, 3, 5, 7, etc.) until it reaches the bottom of the image, whereupon it returns to the top of the image to scan the even-numbered lines (2, 4, 6, 8, etc.). Thus, each scan, or field, comprises only half of the total number of scanning lines, and two fields are required to produce the 525-line frame. Each field is completed in one-half the frame time.

In addition to the picture information, blanking and synchronizing signals are transmitted by the television station to control the intensity and movement of the scanning beam in the picture tube of the television receiver. Both these signals are in the form of rectangular pulses. Moreover, their polarity and amplitude are such that they are received as "black" signals and therefore do not appear on the receiver screen.

The signal bandwidth of NTSC video is 4.2 MHz and is often specified as having about 330 lines of horizontal resolution. Horizontal resolution in TV lines specifically refers to the maximum number of black *and* white elements that can be viewed across 3/4 the width of the screen. In other words, the number of displayed pixels is equal to the horizontal resolution multiplied by 4/3. Thus the number of horizontal pixels representable in standard NTSC is about 440. In the vertical dimension, about 480 of the total 525 scanning lines are visible, so it is tempting to say that the pixel resolution capability of NTSC is 440 (H) by 480 (V). However, the interlace technique of scanning reduces the effective vertical resolution. It is generally accepted that the resolution of an interlaced scanning system is about .7 times the number of scanning lines so the NTSC practical vertical resolution is closer to 340 lines.

Television broadcasting began with monochrome signals. The monochrome television signal is amplitude modulated on an RF carrier, and much of the lower sideband is removed, hence the term vestigial sideband modulation is used. Audio information is frequency modulated on a carrier 4.5 MHz above the visual carrier.

Color is a dimension that was added skillfully to black-and-white television. Color television was authorized by the FCC in 1953 as a culmination of the work of the National Television System Committee. In the NTSC system, the color information is carried as simultaneous amplitude and phase modulation of a 3.58 MHz subcarrier which is specifically related to the scanning rate to avoid visible artifacts and promote better color separation. A diagram of the television channel showing portions occupied by color and monochrome signal components is shown in Figure 2.5.

PAL and SECAM

At the Eleventh Plenary Assembly of the CCIR, held in Oslo in 1966, an important international effort was made at standardization of color television systems by the contributing countries of the world. The discussions pertaining to the possibility of a universal system proved inconclusive. Therefore, the CCIR, instead of issuing a unanimous recommendation for a single system, was forced to issue only a report describing the characteristics and recommendations for a

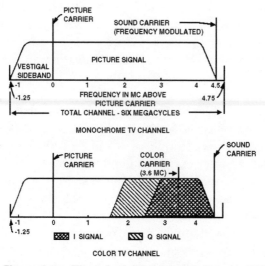

PICTURE
CARRIER

SOUND CARRIER
(FREQUENCY MODULATED)

PICTURE SIGNAL

VESTIGAL
SIDEBAND

-1   0   1   2   3   4   4.5
-1.25   FREQUENCY IN MC ABOVE   4.75
PICTURE CARRIER
TOTAL CHANNEL - SIX MEGACYCLES

MONOCHROME TV CHANNEL

PICTURE
CARRIER

COLOR
CARRIER
(3.6 MC)

SOUND
CARRIER

-1   0   1   2   3   4
-1.25

 I SIGNAL   Q SIGNAL

COLOR TV CHANNEL

Figure 2.5   The television channel. *Source: NAB Engineering Handbook,* **8th Edition.**

variety of proposed systems. It was, therefore, left to the controlling organizations of the individual countries to make their own choice as to which standard to adopt.

This outcome was not totally surprising since one of the primary requirements for any color television system is compatibility with a coexisting monochrome system. In many cases, the monochrome standards already existed and were dictated by such factors as local power line frequencies (relevant to field and frame rates) as well as radio frequency channel allocations and pertinent telecommunications agreements.

Thus, technical factors such as line number, field rate, video bandwidth, modulation technique, and sound carrier frequencies were predetermined and varied in many regions of the world. The countries of Europe delayed the adoption of a color television system, and in the years between 1953 and 1967, a number of alternative systems that were compatible with the 625-line, 50-field existing monochrome systems were devised. The development of these systems was to some extent influenced by the fact that the technology necessary to implement some of the NTSC requirements was still in its infancy. Thus, many of the differences between the

NTSC and other systems are due to technological rather than fundamental theoretical considerations.

Most of the basic techniques of NTSC are incorporated into the two other system approaches, SECAM and PAL. SECAM (Séquentiel Couleur avec Mémoire, for sequential color with memory) was developed and officially adopted by France and the USSR, and broadcast service began in France in 1967. The PAL (Phase Alternate Line) system has been adopted by numerous countries, including continental Europe and the United Kingdom. Public broadcasting began in 1967 in Germany and the United Kingdom using two slightly different variants of the PAL system. The ease with which international exchange program material may be accomplished is hampered by these multiple standards and is accomplished at present by means of standards conversion techniques, or "transcoders," with varying degrees of loss in quality.

HDTV

The United States is currently on the verge of establishing a new technical standard for television broadcasting. In the case of both the monochrome television standard in 1941 and the color standard in 1953, an industry committee developed the technical standards, which were then adopted by the FCC. This is also the process being followed for HDTV standardization.

In February, 1987, 58 broadcasting organizations and companies field a petition at the FCC requesting that the Commission initiate a proceeding to explore the issues arising from the introduction of advanced television technologies and their possible impact on the television broadcasting service. The broadcasting organizations were concerned that the alternative media would be able to deliver HDTV to the viewing public placing terrestrial broadcasting at a severe disadvantage.

The FCC agreed that this was a subject of utmost importance and initiated a proceeding to consider the technical and public policy issues of ATV. In November, 1987 the FCC formed the Advisory Committee on Advanced Television Service, which then expeditiously began its work. Following an initial competitive phase, the remaining proponents in the FCC Advisory Committee process agreed to merge their systems into a single hybrid proposal in May, 1993. This system, currently under consideration by the FCC Advisory Committee, is called the "Grand Alliance." The seven Grand Alliance partners are AT&T, David Sarnoff Research Center, General Instrument Corp., MIT, North American Philips, Thomson Consumer Electronics, and Zenith Electronics.

The features of the Grand Alliance system are as follows:

- Digital video compression based on the International Standards Organization (ISO) MPEG-2 standard.

- A packetized data transport system, allowing the transmission of virtually any combination of video, audio, and data in packets, based on the MPEG-2 transport layer specification.

- Interlaced and progressive scanning formats. The formats supported are 24, 30, and 60 frame-per-second progressive scan with a pixel format of 1280 × 720 (number of active picture elements per line by number of active lines) and 24 and 30 frame-per-second progressive scan with a pixel format of 1920 × 1080. The system also supports 60 frame-per-second interlaced scan with a pixel format of 1920 × 1080.

- Digital audio surround sound based on the 5.1 channel Dolby AC-3 technology.

- Rugged digital transmission technology for robust delivery of broadcast signals.

The current schedule anticipated a recommendation on a standard to the FCC by the Advisory Committee no sooner than the spring of 1995.

MPEG

Digital compression of video and audio signals is certain to be an important part of the future universe of broadcasting. One of the most important groups involved in setting international standards for digital video and audio compression is MPEG.

The Moving Picture Experts Group (MPEG) is a joint committee of the International Organization for Standardization (ISO) and the International Electrotechnical Commission (IEC). The charter of the MPEG video and audio groups is to develop compression standards for full-motion video, associated audio, and their multiplex for digital storage media.

The original MPEG-1 video and audio standards were nominally developed for 30 frames/second progressive scan, low resolution video (352H × 240V) at 1.5 Mbits/sec data rate, and stereo audio at a 256 kbits/sec data rate. Header/descriptors incorporated in the standards, however, allow modification of the nominal parameters, including changes in picture size, resolution and aspect ratio, pixel aspect ratio, frame rate, and compressed data rate. The MPEG-1 standards were officially adopted by ISO in 1992. MPEG-1 is often described as "VHS tape quality."

Ongoing work in the MPEG committees is developing an MPEG-2 video and audio compression standard, which is intended to include broadcast quality and HDTV applications. MPEG-2 work began in 1990 with the initial intent of developing a generic standard that could be used for all applications. However, an all-things-to-all-applications algorithm was eventually abandoned in favor of developing a "toolkit" of related techniques, encompassing the concepts of *profiles* and *levels*. A *profile* is a defined subset of the entire bit stream syntax. A *level* is a defined set of constraints imposed on parameters in the bit stream.

The draft MPEG-2 video standard is expected to achieve full status as an International Standard in November, 1994. While the audio standard for MPEG is also in draft status, audio standards for MPEG-2 will not likely be finished until

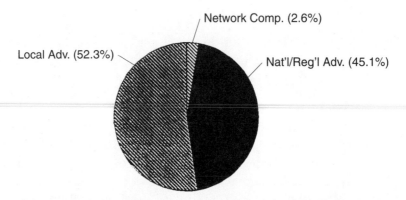

Figure 2.6   1992 national average station revenue. *Source:* 1993 TV Financial Report, NAB.

March 1997, due to recent positive consideration of non-backward compatible coding modes (coding that is not compatible with MPEG-1 audio).

Infrastructure

Expenses and Revenue for a Broadcast Station

Television broadcasting depends on the selling of advertisements as the major source of station revenue. Figure 2.6 shows average television station revenue sources.

In 1992, 45.1% came from national/regional advertising and 52.3% came from local advertising. Network compensation accounted for 2.6% of revenue. The networks pay for their affiliates to broadcast network programming. This is how the network is created, and it assures the network of clearance not only for its programs but also for network commercials.

Expense sources for an average television station in 1992 are shown in Figure 2.7.

The largest cost item for an average television station is programming and production. The station must buy or produce all of its programming. G&A is also a large expense, covering such functions as accounting, office management, traffic, and public and community affairs.

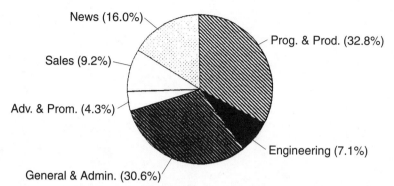

News (16.0%)

Sales (9.2%)

Adv. & Prom. (4.3%)

General & Admin. (30.6%)

Prog. & Prod. (32.8%)

Engineering (7.1%)

Figure 2.7   1992 national average station expenses. *Source:* **1993 TV Financial Report, NAB.**

Local television stations (not the networks) spend approximately $4.5 billion annually securing the right to air a broadcast program. This comes from an average figure of $2.2 million for a network affiliate and $6.8 million for an independent. The television networks, taken together, spend close to $3.5 billion on program development, $6.5 billion in total programming expenses, including acquisition of program rights and license fees.

The costs of producing a television program can become quite complicated. As an example, however, an average first-run sitcom (i.e., with no track record of an audience) costs upwards of $750,000 to produce. The owner of the program, usually a Hollywood producer, sells a license fee to the broadcast network. When the network airs a broadcast, it does not "own" the program; it has merely paid for the right to air it. These license fees are highly variable, but may run $500,000 to $900,000 per episode. The studio or program owner will try to recoup additional revenue through selling the program as well to cable, rerun, or independent markets. The network pays a premium for first-run rights. The economics of this depend a lot on the success of the program.

The profitability of the television business depends on a number of factors, a major contributor being the local market conditions. By and large, television station profits have been declining over the past few years. In 1991, the median pretax profit for an independent (i.e., not affiliated with a network) local television station was actually a $300,000 loss.

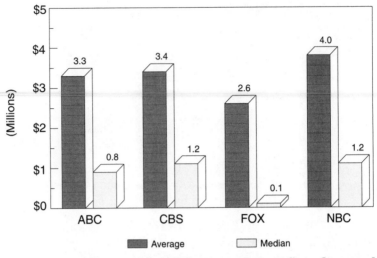

Figure 2.8 1992 network affiliates pretax profits. _Source:_ A Profile of Television Stations by Network Affiliation, NAB.

Average and median pretax profits for network affiliate stations in 1992 are shown in Figure 2.8. The relationship between the average and median values (averages being significantly higher than the medians) imply that several stations affiliated with each of the networks made substantially more than many of their network's other affiliates. This is seen more clearly when the different market size values are examined. Table 2.1 shows net revenue and Table 2.2 shows pretax profits as a function of ADI (Area of Dominant Influence). The decrease in both revenue and profit in smaller markets is quite apparent.

Impact of HDTV on Broadcasters

The financial impact on stations of completely converting to HDTV (or generally, ATV) service is daunting. However, the planning and scheduling of the necessary capital acquisitions and their installation will permit the costs to be spread over a number of years, while meeting the regulatory requirements expected.

Network "pass-through," or its equivalent for syndicated programming, will be the prevalent service of ATV stations

Table 2.1 **1992 Average Network Affiliate Net Revenues versus Market Size.** *Source:* **A Profile of Television Stations by Network Affiliation, NAB.**

ADI Range	ABC (000s)	CBS (000s)	FOX (000s)	NBC (000s)
1–25	$61,019	$50,932	$41,827	$56,197
26–50	18,247	17,068	11,308	17,889
51–75	10,265	10,582	6,166	10,623
76–100	5,833	7,747	3,911	8,500
101+	4,229	4,391	2,302	4,367

at the start. Pass-through essentially means receiving signals from a program source such as a network or other distributor and passing them through the television station with little or no change. For commercial stations, the ability to play back commercials is likely to be a necessity from the beginning, if there is to be any revenue generated by the operation. This may be a modest capability at the start, with increased capacity installed as the audience and revenues from ATV increase. NTSC programs "upconverted" to the ATV transmission format are also likely to be prevalent in the early years of the new service.

To assess the cost of ATV, a "transitional" station has been defined. This is a pass-through operation that provides the

Table 2.2 **1992 Average Network Affiliate Pretax Profits versus Market Size.** *Source:* **A Profile of Television Stations by Network Affiliation, NAB.**

ADI Range	ABC (000s)	CBS (000s)	FOX (000s)	NBC (000s)
1–25	$17,636	$15,069	$11,963	$20,185
26–50	2,377	3,239	961	3,974
51–75	1,711	2,008	413	891
76–100	553	2,014	(90)	679
101+	91	(307)	(275)	264

capability for most of the production techniques that are applied to program continuity integration today. Thus it will allow a moderate amount of postproduction in the station's release operation. It includes the ability to record and play back programs, commercials, and promos. It includes the ability to generate graphics of HD quality. It includes the ability to superimpose messages and images in the same fashion as is currently done in typical NTSC operations. The estimated cost of such a transitional station ranges from $1.9 million to $2.2 million with best-case assumptions.

Incremental revenues from ATV programming can be gained only if there is a sufficient number of TV households with ATV equipment. The prediction that viewers will buy this equipment is based on the supposition that the greatly superior quality of the HD picture and sound will be compelling to viewers of programs in which image quality, resolution, the wide screen, and CD quality sound are important factors. These programs might include feature films, where for the first time, the attributes of the cinema experience may be brought into the living room, possibly saving the expense of "going to the movies." TV movies and miniseries are other categories that may attract the discerning viewer to HDTV. The merits of HDTV are perhaps of the greatest significance in the coverage of sports. The wide screen and the high resolution afford the viewer a more complete appreciation of the game, permitting wider-angle coverage, and creating a sense of actual presence at the game.

Commercials inserted in such high-profile programs could perhaps be marketed at a premium rate, securing additional revenue. Also, high-profile programs that are simulcast may be broadcast at different times on the NTSC and ATV channels, effectively doubling the market exposure, and increasing the potential viewership, and hence revenues.

One of the benefits of the use of digital transmission in the ATV service is that there promises to be capacity left over beyond that needed to carry the basic video and audio of television programming. This capacity can be put to additional uses, some of which may bring revenue that can help pay for the extra costs entailed by the addition of a second channel and all it implies. Examples of the kinds of data transmission services that may become possible are the delivery of financial data, the delivery of news and other information, the delivery of computer programs and games, publication of magazines, and other information delivery services. It may be possible to send alternate video and audio program services. This capacity is likely to increase with time as the compression methods for both the main programming service and for alternate programming services improve.

Regulation and Legal Constraints

FCC Rules and Regulations

Broadcasting is subject to a multiplicity of laws and regulations from various, and sometimes conflicting, jurisdictions. The centerpiece of broadcast regulation is the Communications Act of 1934, which is administered by the Federal Communications Commission. The FCC allocates spectrum for services, assigns frequencies to individual stations within the allocated spectrum, licenses new stations, and regulates existing stations. Regulation includes inspection for compliance with FCC rules and technical standards and processing renewal of licenses. Broadcast stations are licensed to serve the public interest, convenience, and necessity. At the time of license renewal (maximum of every five years for television stations), the station's record in this regard is examined. Failure to comply with the Commission rules results in reprimands, fines, denial of license renewal, or license revocation, depending on the seriousness of the violation.

RF Spectrum Allocation for Terrestrial Broadcasting

There are several frequency bands used for television broadcasting within the United States. The plan includes the low-VHF band TV channels 2 to 4 (54 MHz to 72 MHz) as well as channels 5 and 6 (76 MHz to 88 MHz), the high-VHF band channels 7 through 13 (174 MHz to 216 MHz), and the UHF band channels 14 through 69 (470 MHz through 806 MHz). Each television channel is 6 MHz wide.

Television also uses RF spectrum for auxiliary use in support of the broadcasting service. There are several types of television broadcast auxiliary stations: TV pickup stations, TV STL stations (studio transmitter link), TV relay stations, TV translator relay stations, and TV microwave booster stations. TV pickup stations are land mobile stations used for the transmission of TV program material and related communications from the scenes of events to TV broadcast or low-power TV (LPTV) stations. TV STL stations are fixed stations used for the transmission of program material and related communications from the studio to the transmitter of a TV broadcast or LPTV station. TV relay stations are fixed stations used for transmitting visual program material between TV broadcast or LPTV stations or for the relay of transmissions from a remote pickup station to a single TV station. TV translator relay stations are fixed stations used for relaying programs and signals of TV broadcast stations to LPTV stations or TV translator stations. TV microwave booster stations are fixed stations used to receive and amplify signals of TV pickup, TV STL, TV relay, or TV translator relay stations and retransmit them on the same frequency. These stations are used to transmit signals over a path that cannot be covered by a single transmitter.

TV auxiliary channels are available for operation in a handful of allocated frequency bands between 1990 MHz and 40 GHz.

Station Ownership Regulations

The early days of broadcasting were marked by the individual entrepreneur who owned a station in a particular community and remained associated with that facility for many years. As broadcasting grew, there emerged a pattern of multiple ownership—ownership by nonbroadcast entities such as newspapers, and ownership by networks. The Commission expressed concern with placing too many "voices" under a single control and initiated regulations to control that concern. Specifically, no person or group may own more than 12 television stations. Also, TV stations under common ownership may not operate in markets collectively containing more than 25% of the nation's TV homes. Common ownership of TV stations and newspapers or TV stations and cable in the same market is also prohibited, although some relaxation of ownership rules is being considered by the Commission.

Decency and Children's TV

Defining indecency is somewhat qualitative: the Commission considers broadcast material to be indecent if it describes, in terms patently offensive as measured by contemporary community standards for the broadcast medium, sexual or excretory activities and organs, at times of the day when there is a reasonable risk that children may be in the audience. The seriousness of the Commission's concern with indecent program is apparent from the fact that it has levied sizable fines on stations alleged to have violated the Commission's rules against indecency.

In 1991, the FCC implemented the Children's Television Act of 1990. This requires that each television broadcast station is required to serve the "educational and informational needs of children" in its overall programming. In addition, during commercial programs that are intended for an audience of children 12 years and under, commercials can take up a maximum of 10 1/2 minutes per hour on weekends and 12 minutes per hour during the week.

HDTV Requirements

The FCC has been considering the transition to ATV broadcasting since 1987. Some of the FCC decisions that have been made since that time that are of particular interest are shown below.

Only existing broadcasters will be initially eligible for the assignment of ATV frequencies.

Broadcasters will have three years from the effective date of the selection of an ATV system, or the issuance of a final Table of ATV Frequency Allotments, whichever is later, in which to apply for an ATV channel and construction permit. (The baseline date when application may first be made, is denoted as Year 0.)

Broadcasters will have a total of six years from Year 0 in which to construct the ATV facility and have it ready for service. Thus, the earlier a broadcast station applies for a license in the period Year 0 to Year 3, the longer it will have to construct the ATV facility. At the end of the six-year period, stations are required to transmit an ATV signal. Broadcasters failing to meet these deadlines would forfeit their initial exclusive eligibility for a set-aside channel, but would still remain eligible to apply for an ATV channel at a later date.

In Year 7, broadcasters must implement 50% simulcasting on the paired NTSC-ATV channels of the station. In Year 9, broadcasters must implement 100% simulcasting. The FCC has defined simulcasting as the broadcast on the NTSC channel of the same basic material broadcasting on the ATV channel, excluding commercials and promotions, within a 24-hour period.

The FCC has emphasized that ATV is a replacement service for NTSC and, therefore, as soon as ATV service predominates in the marketplace, NTSC service will be terminated and one of the two 6 MHz channels operated by broadcasters will be returned to the FCC for other uses. As a preliminary matter, the FCC has established Year 15 as a target schedule for when broadcasters must convert fully to ATV service. At that time, all NTSC broadcasts would cease. Broadcasters who had not converted to ATV service by that date would have to cease broadcasting.

Outlook

The future outlook for television broadcasting is very positive. There is no indication that there has been any decline in the public's demand for news or entertainment. Through the introduction of video games, the expansion of the home video industry and the explosion of home computer usage, the average number of hours per week spent viewing television has held nearly constant for more than two decades. There is no indication that this pattern is likely to change, and there is some indication that it may expand as people practice higher levels of "cocooning" during the "threatening" nineties. The challenge for broadcasters is to retain their dominant position as the suppliers of programming and to protect the viability of the delivery conduit that they presently possess directly to consumers.

It is the flexibility and universality of this "conduit" that is broadcasting's most powerful advantage. No other medium can deliver the same amount of information to the same number of people in such widely diverse locations at such low cost. Broadcasters are currently fighting to preserve this "right" as new technologies emerge. Serious decisions regarding what, if any, limitations should be imposed on existing broadcasters as they develop new digitally based services will determine how effective those services may be; what they will cost the public; and how competitive broadcast delivery will be compared to broadband cable, common carrier telephone, direct broadcast satellite, low-orbit earth satellites, narrow area cellular, and other alternative choices for service delivery.

The strategy being pursued by the traditional broadcast networks indicates their understanding of this situation. While they continue to aggressively develop new programming to distribute through their affiliated stations, they are also developing complementary businesses to broaden their revenue base. Take as an example recent projects by the National Broadcasting Company (NBC). They are currently operating a full-time news and information channel for cable delivery (CNBC), cooperating with America On-Line in disseminating a computer-based news service, and working with Ziff-Davis Interactive services to develop a high-speed "Data Casting" service that will be delivered to consumers' desktop computers through a combination of satellite and terrestrial broadcasting station subcarriers. They are also providing both contemporary and library video programs to telephone companies who are experimenting with "Video-on-Demand" in New York City and other locations.

The American Broadcasting Company (ABC) is pursuing similar strategies with on-line services and cable. This program is supported by a strong international marketing effort for their program and news services. This international market is being served by a combination of satellite and cable delivery with additional over-the-air service planned as developing nations make the facilities available.

While some broadcasters have expressed concern about the networks' "nontraditional" business enterprises, most are comfortable with the service their network provides and feel that they will continue to have a mutually beneficial business relationship in the future. The local affiliates continue to push the networks to develop data products that they can deliver to their regions, and many are looking to add local content to these data streams as they deliver them to their customers.

These efforts are being given a boost by the National Data Broadcasting Committee, a cooperative effort of the National Association of Broadcasters and the Electronic Industries Association. This group's primary goal is to foster establishment of a national technical standard for high-speed data services on existing and future broadcast services. This type of effort will preserve the broadcasters' ability to serve the public as the demand for selectivity and diversity continue to grow.

The desire for the ability to select what is consumed seems to be the most powerful element in the public's interest in new and expanded media. This is not a new trend but simply the continuation of a long-term evolution. This is governed by two basic principles. The first is that once a consumer is given a service or ability, a new service must supply at least that level of ability before it will be accepted. In other words, you can't take away something that people are accustomed to having. The second principle is that people will always desire more control over what they elect to receive. These two principles continue to come into play as media planners and advertisers experiment to determine how both programming and advertising will be tailored to take advantage of the enhanced capabilities of evolving "new media."

The first principle seems to apply most strongly in programming. We are rapidly approaching a time when people could actively participate in TV programming in a variety of ways. These include playing along with quiz shows, selecting their own camera shots at sporting events, registering "instant feedback" during a political debate, or requesting additional information or placing an order during a commercial. In early trials, a vast majority of users made very limited use of these abilities. Researchers are debating the cause of this situation. Some theorize that people have become accustomed to using incoming media programming in a passive way and simply want to continue being informed or entertained without being forced to react or interact. An example is the usage of video music channels during the Bell Atlantic employee trials in northern Virginia. Along with a number of other services, these test users were given the ability to select "custom menus" of music videos to play starting at a selected time. After one or two initial tries, few people used this service regularly. The menu was modified to offer a "system-selected custom sequence" as an option. The use of the service increased noticeably and held at the newly elevated level. People apparently were used to having their programming selected for them and were not willing to invest the time to make their own selections except for "special" occasions. There is another interpretation of these results. Some observers feel that failure to use interactive features is due to lack of training in their use or to "unfriendly" design of the user interface. The proof of either theory will be what usage patterns emerge in the long term.

One result consistently emerges as new media are offered: people want the freedom to choose the time and location to consume certain key programs. This phenomenon began with the use of home VCRs to time-shift programs to suit viewer schedules and continues as various video-on-demand options are tested. This "selective consumption" tends to be used by higher income, higher educated consumers and less by traditional "blue collar" consumers. This desire for the ability to suit delivery time to need, as opposed to a wider selection of choices or the ability to support interactive service enhancements, seems to be the greatest factor in creating usage for early "video-on-demand" trials.

This desire for choice is the vehicle that most advertisers will ride into the information age. The advertising industry has been deeply concerned about how to react to the dilution of the impact of a given advertisement as audiences were splintered by the increased number of viewing choices. Except during certain keystone events, it is no longer possible to deliver the same message to 60% of the country at the same time by buying time on "the big three" networks. First attempts to counter this trend found advertisers struggling to guess which consumers would watch which options and then producing a wide number of targeted spots for these selected viewers. This process is expensive and uncertain. The promise of the future is the ability to use expanded technical options to determine who is using a given form of advertising and then develop products, services, and information packages that closely match the choices favored by that group.

Once again, early trials of interactive systems are supporting the validity of this approach. Users do seek the information contained in product advertising. They will watch and respond to programming that delivers information about products or services that are of use to them. The secret for advertisers is to use the feedback they receive on viewers to build a database that directs their future marketing efforts.

The "new consumers" also respond to subtle product placement in entertainment programming. One trial that placed advanced consumer equipment in the set of an adventure show, with the option of "clicking" on the objects to receive more information about them, proved highly successful. People will voluntarily seek "catalogue" type information on products or services when a group of selections is made available.

Another interesting result from video trials is that people appear very willing to trade watching several minutes of commercial content for a waiver or reduction in the service fee for video on demand. This is encouraging affirmation of the continuing viability of the traditional advertiser supported service that has been the mainstay of mass market, widely distributed broadcast services. The challenge for broadcasters is to develop appropriate technologies to direct the desired information to the segment of the audience that needs it—in other words, to use new technology to tailor the service to the user.

These options may seem to imply the need for a radically different system of distribution. This is simply not the case. By using technologies which are now available, broadcasters have a great advantage over other media. They have the bandwidth or capacity to deliver enormous amounts of information quickly and dependably over a wide area. Using emerging digital technology, they can add the custom interactive features that allow people the ability to make choices in their viewing selections and take advantage of response opportunities that might be offered. No other medium can offer equivalent economies of scale.

A powerful advantage of broadcasting for the future is that it is wireless. Wireless broadcast transmissions provide near universal coverage. Broadcasting is the closest to achieving universal coverage of any current wireline or wireless provider with 98% of households having a TV set. An obvious benefit of wireless transmissions is that they can be where the customer is rather than the customer needing to be where the service terminates. The benefits to consumers who use laptop computers, personal digital assistants, or portable televisions are obvious. Wireless transmission is also inexpensive and is the most efficient form of transmission. There is no need for expensive fiber, coaxial, or copper cable. Because wireless transmission systems do not require these heavy capital investments, they are less costly in the short and long run. Wireless transmissions can also involve a mix of support mechanisms. Wireless can be totally free: it can be advertiser supported; it can be addressable for subscription fees; or it can be encrypted for specialized rates. It is very unlikely that any wireline provider can meet the standards of wireless.

Perhaps the most important advantage of broadcasters for the future is experience. Broadcasters have been the primary source of local news and information programming in American communities for the greater part of the twentieth century. They are experts in determining the needs and concerns of the community and delivering programming services that people want to use. Whether it is delivered directly over the air, relayed by cable, stored and forwarded on a digital network, or beamed by direct satellite, the vast majority of the programming people want to watch was or is generated by broadcasters. This expertise in getting product to market is a powerful force. This capability combined with the expanding technical possibilities establishes broadcasters as the most powerful players and best potential partners for "New Media" in the nineties.

The National Association of Broadcasters

The National Association of Broadcasters (NAB), founded in 1923, is a not-for-profit trade association representing the nation's radio and television broadcast stations and networks. NAB has three major functions: representation of the broadcasting industry in Washington, services to NAB members, and industry conference and exhibition management. NAB has over 150 employees and an annual operating budget of approximately $19 million. The staff tracks and manages over 150 major issues affecting broadcasters ranging from tax law and environmental impacts to children's television issues and many technical issues, such as high-definition television and interactive multimedia.

NAB's members are local radio stations and local television stations. The membership structure reflects the local nature of the United States broadcasting industry. The FCC licenses individual stations to operate in local communities in the public interest. Each station determines for itself how best to program its station—how much local news, how much nationally cleared programming, and how many movies. Broadcast networks are also members of NAB, but these are not licensed by the FCC. As it is, only the stations that use radio spectrum for broadcasting are licensed.

The NAB Spring Conference and Exhibition is perhaps NAB's best-known activity, consisting of over 800 exhibitors and 250 conference sessions with attendance over 70,000. NAB held its first annual convention in October 1923. Station representatives attending the convention discussed issues of vital importance to broadcasters such as music licensing and the impending regulation by the Federal Radio Commission. Today's NAB Spring Conference and Exhibition consists of conference sessions that teach and inform attendees about the most useful information that can be obtained. It is also intended to create an environment that brings buyers to meet with sellers, to show the newest and most useful technologies, and generally to facilitate the conduct of business in a fast changing and technologically advanced marketplace.

3

CABLE Television: Views on the Set-Top Box

WALTER S. CICIORA

In the early 1950s, the cable industry was formed mainly to supply remote areas unable to receive clear broadcast television signals due to distance or physical obstructions. The cable industry has, of course, grown in many new directions since that time and has now become a formidable transport competitor to over-the-air broadcast through the introduction of new channels, content, and other services. Nonetheless, the cable industry's historical ties to broadcast must not be ignored and will result in significant impact on the design of so-called set-top devices.

The original design and purpose of set-top devices are different than those of the much-hyped interactive digital services. These devices were converters that overcame physical limitations of the prevailing consumer devices, dealt with troublesome interference with broadcast signals, and provided protection from signal theft (to varying degrees). These issues were, and still are, near and dear to cable providers since they impact their business directly. Until the nation converts from traditional NTSC television sets to the next-generation digital set, set-top designers will have to continue to wrestle with these issues and will have to do so in the analog domain.

Government regulation came to the cable industry later than it did for broadcast television, grappling with cable industry-specific issues such as signal leakage and, more recently, standardization of "cable ready" television sets. While both the subject and amount of government regulation differs from broadcast television, the oversight of this industry has become progressively tighter. One can assume that the regulatory environment for broadcast and cable will eventually be harmonized as the differences between the industries continue to be redefined. In one respect they must converge: they both target television receivers and hence must coexist, even if uncomfortably at times.

The cable industry has different concerns from other industries. This chapter gives a good view of security issues that are less likely to be of concern to broadcasters: subscriber privacy and signal interdiction. The privacy issue conjures images of big brother monitoring the individual viewing habits of consumers and drawing conclusions from such behavior. Blocking signals except to those validated to be legitimate paying customers is, of course, an economic concern. Both topics are bigger issues to the cable industry than many realize.

Finally, just because a cable system uses wires, don't assume that it is just like a telephone system. In fact, it is built to order to provide a specific kind of service and is not easily changed or recycled to offer, say, telecommunications services. This chapter reviews some of the architectural differences that form the basis of substantial capital investments. These investments are of such a scale that the implementation of a Class 5 all-digital system (see Chapter 1), instead of a lower technology system, are extremely complex.

The historical elements of this industry once again play a huge role in shaping the economics of the business and in some ways retard the movement to newer technologies. Yet the common genetic material that makes up cable and broadcast forces them to move in lockstep when the transmission medium is redefined from analog to digital. For, in the end, cable (and indeed direct broadcast digital satellite) must always convert to a broadcast-compatible signal in order to connect to and be compatible with television sets.

Introduction

The term "set-top box" is inadequate and increasingly inappropriate for the Broadband Platform, which is being created to bring a new range of exciting services to the subscriber. While this is recognized, a fitting substitute has not come about. With some hesitation, "set-top box" will be used in this paper, but only because better terminology has not yet been found.

There is a great deal of attention being paid to the set-top box in the last couple of years. There are those who are trying to eliminate it because they see it as getting in the way of products they would like to sell. There are others who wish they could sell it. Still others see it as a potential gateway which may limit their access to consumers. Finally, there are those who see it as a wonderful facilitator, which makes possible new and innovative services without risking the subscribers' hardware investment.

Early History and TV Tuner Deficiencies

In the beginning, cable systems delivered just a few channels, fewer than the twelve the common TV receiver could accept. There were no VCRs. Cable was the "Community Antenna Television" (CATV) service providing TV where few high-quality signals were available because of over-the-air propagation problems. The first set-top boxes were designed not to give access to more channels but to compensate for a TV set deficiency when used on cable. That deficiency is called Direct Pick Up (DPU) interference. DPU means that the TV's internal circuits are not adequately shielded from electrical signals. They directly pick up these signals from the environment. In the broadcast application, this is no problem. The antenna signal and the DPU signal are the same and they simply add and provide more signal for the TV. In the cable situation, there is a problem. If the cable channel is different from the off-air channel, the two signals will mix inside the TV, resulting in interference. If the DPU is mild, the picture is distorted and the result is annoying. If the DPU is strong, the result may be unusable. Even in the case where the cable channel is the same channel as the off-air channel, a problem arises. This is because the speed of propagation for a signal through a coaxial cable is about two-thirds the speed through the air. So the DPU signal gets to the TV before the cable signal. The time difference is such that the DPU signal appears on the screen to the left of the cable signal, generating a "ghost" image. Again, in mild cases, that ghost could be just annoying; in strong cases, it can completely spoil the picture. Cable channels also exist at frequencies used by other strong signals. Cable channels 19 through 21 are frequently troubled by transmission from pager services.

To overcome DPU, early cable systems introduced the set-top converter. It had a very well-shielded tuner so that it suffered no DPU. It then *converted* the off-air channel to one of the channels not used locally, usually channel 3 or 4. Since the output channel of the converter was not used locally, there was no signal to directly pick up and the DPU problem was solved. The invention was covered by the Mandell patent and had nothing to do with expansion of channel capacity.

DPU is not the only TV receiver deficiency which neutralized with the use of a set-top when connected to cable. TV receivers are designed for use with broadcast signals. They perform that task very well. The technical requirements for off-air reception have two characteristics. First, the FCC almost never assigns broadcasters to adjacent channels in a local market. There almost always are one or two blank channels between occupied off-air channels. Secondly, the tuner must be able to pick out a weak desired channel from between strong undesired channels. So if there is a ball game in the next town on a channel that is between a couple of strong local channels, the sports fan can pick it out and watch it. The cable television application is different because it filled up all the channels, leaving no gaps. However, the signal strengths of the channels were nearly identical. The first characteristic makes the job more difficult, while the second makes it easier. Both together made it different. Many TV tuners go into "overload" when presented with a full cable spectrum. The signals become distorted and moving diagonal patterns of interference are visible in the background in mild cases. More severe cases result in loss of channels and service calls from the cable operator to "fix" this problem. The only solution is to either attenuate the cable signals, which yields snowy pictures, or to put on a set-top box.

TV tuners are designed for competitive advantage on the salesroom floor when viewing broadcast stations. Since most TV stores don't have cable, the consumer cannot evaluate cable service performance until taking the set home. Then it is too late.

There is an interesting failure of logic with the DPU problem. A TV set with DPU generally performs very well when connected to an external antenna. When connected to a cable, the result is less than satisfactory. Simpleminded, first-order logic has it that the cable must be at fault. The cable system gets the complaint, not the TV manufacturer. The cable technician installs a set-top with a superior, more expensive tuner, and the problem is solved. Frequently, the subscriber thinks it is all part of some sort of cable industry plot to gain a monthly rental fee.

Unfortunately, DPU and other tuner deficiencies remain problems today. This is because their complete solution usually requires a "double conversion" tuner, which is about twice as expensive as the type used in all TVs and VCRs to date. Since the consumer electronics market is so competitive, that extra cost is not included in today's products. No consumer products made today include a double conversion tuner.

TV receivers generate electrical signals in the process of creating pictures. Modern TVs with digital tuning and On-

Screen Displays (OSDs) generate digital signals, which behave just like radio signals. Those signals must be kept inside the TV or they will cause interference with others' reception. If conducted to another TV in the home or in a neighbor's home, interfering patterns may appear on the screen. The set-top box very effectively isolates the TV from the cable system and protects the reception of others.

There is another occasional function performed by the set top: protection from lightning surges. A common experience with lightning strikes: the $150 TV set with a set-top box was undamaged, although the set top was replaced by the cable company, while the $500 "cable-ready" TV cost $40 to find out that it was beyond repair.

Cable vs. Telephone CPE

We frequently hear comparisons between the telephone industry and the cable industry as justifications for similar handling of Customer Premise Equipment (CPE). It is pointed out that at one time the phone company had a monopoly on CPE and did not allow any "foreign" equipment to be used on its lines. The phone company used the argument that it must protect its network. "Foreign" equipment would surely damage it to the detriment of all subscribers. Of course, no such disaster happened. In fact, the popular press and even the trade press have no reports of such damage. On the contrary, a market for innovative CPE flourished, and consumers benefited.

Moreover, unlike telephone systems, cable TV systems do not conform to any national standard, and the technology can vary from one cable system to another. A set-top descrambler used in one cable system probably won't work in another system. But this promotes innovation and diversity in cable TV technology, while telephone industry standards stifle innovation.

The temptation is great to extend this thinking to the cable industry by way of analogy. The cable industry has been accused of being a monopoly and of wanting a monopoly position in CPE. The claim is made that the cable industry's concern about damage to its network is likewise a red herring. All this serves to point out the weaknesses of thinking by analogy. There *are* fundamental differences between the telephone service and cable service which require some careful analysis. Careless application of what worked with the phone company to the cable company is liable to generate a variety of unpleasant unintended consequences.

There are real differences between cable and telephone technology which give rise to genuine hazards. The signal frequencies used in telephony do not interfere with other services if they leak out of the telephone plant. The same is not true about cable signals. The FCC has recognized some of these hazards and has created rules relating to the leakage of cable signals into the environment where they may interfere with users of the radio spectrum. Some of this interference is merely an annoyance, while other interference constitutes a serious safety hazard. Aircraft navigation and communication are of high concern. When CPE is connected to a cable system, the CPE internal circuitry must be designed so as to maintain the integrity of the cable plant, keeping cable signals inside and other signals outside. If the cable plant carefully contains the potentially interfering signals, but they leak from inadequate shielding in the CPE, a hazard is created which endangers public safety, and other users of the radio spectrum may be blocked. This design requirement is not part of its normal specification when used for over-the-air reception.

The Role of Security

Signal security has two important aspects: (1) privacy of subscribers, and (2) protection of the rights of intellectual property owners.

In the age of passive television, the privacy issue was limited to simply not monitoring who watches which programs. A common example given is that it is no one's business if the minister subscribes to the Playboy Channel. There is a general concern about "Big Brother" collecting statistics on what programming is watched to generate a "profile" for marketing, political, and other purposes. These issues become much more important in the interactive, two-way world. In that environment, transactions occur, money is spent, and private information is passed back and forth. Not only must the credit card number and the checking account balance be kept secure from prying eyes, but other information that is even more sensitive needs protection. The interactive world promises medical, legal, financial, and other consulting to be done over the information superhighway as efficiency and convenience measures. Your social disease, your divorce or suit against your neighbor, and your bankruptcy proceeding are all very private but very interesting to others who might seek advantage by knowing more.

While the rights of privacy of subscribers are generally recognized, less appreciation exists for the importance of protecting the rights of owners of intellectual property. Without compelling programming, there is no need to consider any of the set-top issues at all. Interesting programming will not exist without the work of artists and creative contributors. Their work must be supported and rewarded. Without that stimulus, the well dries up and programming as we know it ceases. To protect intellectual property, there must be a mechanism for controlling access to the signals in the cable plant. If there was no such method, signals would be taken without payment by the unscrupulous. As an apparently "victimless crime," cable service would be stolen by those who might never consider the theft of physical property. Those who pay for subscriptions would be cheated by those who steal. The creators of programming and the holders of copyrights would soon withhold the fruits of their labors, since this mechanism of distribution would deprive them of a livelihood. The method of controlling access to the basic services is the physical connection to the cable plant. If cable service were just a take-it-or-leave-it proposition, this might be adequate protection. However, cable offers a wide variety of services, which are offered individually and in packages. These product offerings change as tastes change and as the bandwidth of cable systems expands. The most effective method of protecting these signals in a modern cable system is scrambling. This was recognized by the Cable Consumer Compatibility Advisory Group (C³AG) to the FCC in its *Supplemental Comments of the Cable Consumer Electronics Compatibility Advisory Group* submitted July 21, 1993:

> In earlier filings in this docket, the consumer electronics industry has advocated use of consumer-friendly anti-theft measures such as traps, interdiction, broadband descrambling, and other "In-the-Clear" approaches. The cable industry, however, has made a persuasive case that, while all of these may have their virtues—and individual cable operators may find them to be appropriate solutions to their particular needs—none of them is suitable for universal deployment; each has limitations and characteristics that prevent it from reasonably being prescribed as a mandatory solution to compatibility issues. The Advisory Group recognizes that scrambling and encryption are an important part of providing cable services and will remain an essential part of delivering video signals.

Cable TV security technology varies in strength and cost from one cable system to another. There is no national standard for signal security. Indeed, it is generally believed that a single national security standard would be defeated because as a standard it would require public disclosure, thereby allowing hackers to focus all their efforts on a single technology. A diversity of security approaches precludes this focus of efforts.

Until recently, the set-top box was just a cable device which protected cable signals. As the "information superhighway" is constructed and put into operation, the important lessons learned from cable experience should be utilized.

The Joint Engineering Committee (JEC) of the Electronic Industries Association (EIA) and the National Cable Television Association (NCTA) has formed a National Renewable Security Interface subcommittee. In its deliberations, it has concluded that it is important for all the electronics involved in signal security—every last transistor!—to be removable from the equipment. It has further been recognized that the signal security electronics must remain the property of the service provider. This is so that it can be replaced if it is compromised without expense to the subscriber.

There is a great temptation to claim that in the digital age, the encryption technology will be so secure that it can be owned by the subscriber. Analogies to bank cards are made. The "Smart Card" is sometimes used as the mechanism for accomplishing this. While all this sounds appealing at first blush, it has some critical flaws. Perhaps the most important problem is this:

> If there is a breach of security, and the consumer owns the electronics for protecting security, and it is built into the consumer electronics hardware, how is the problem solved? Who pays for the solution of the problem?

The only answer given by the proponents of this approach is "Trust me, it won't happen." That is not an adequate answer. Until there is a contingency plan—including a method of paying for the transition—to deal with the replacement of electronics in the event of a security breach, there can be no agreement to allow consumer ownership of the signal security electronics. Even with a contingency and financing plan, it must be recognized that there will be massive consumer inconvenience and resistance. It is much simpler and easier for all if the service provider retains ownership and responsibility for correcting a breach in security.

A second problem with the belief that "digital" makes perfect security possible is the failure to recognize the resourcefulness of the digital hacker. The hacker has available incredibly powerful computing power at very low cost in modern personal computers. In addition, he has the ability to network these computers together to form arbitrarily large systems. The power of these machines grows almost monthly. It would be foolish to claim that a signal's security is invulnerable to attack by such well-equipped adversaries.

Service Differences

There are fundamental differences in the services offered to subscribers by telephone companies and cable companies. These differences impact on the nature of the in-home hardware used for those services.

Telephone service involves a unique connection between two points, facilitating communications between people or equipment at those locations. Usually the communication consists of two-way symmetric interaction. The conference call is a simple extension of this principle. There are two technical compatibility issues involved in these applications. First, the CPE must be compatible with the network. It must not damage it or be damaged by it, and it must function within its technical requirements. Secondly, the CPE used by both sides of a particular telephone connection must work with each other. They do not have to work with any other equipment, just with each other! That is a major difference between telephony and cable. As an example, two hobbyists or experimenters can devise a pair of unique CPE devices and interact with each other as long as the devices are compatible with the network. They do not have to be compatible with any other CPE hardware. There are probably requirements to get the equipment approved and registered, but that is a matter of form rather than substance. While this type of application is the exception, it demonstrates an important point about telephone communications.

Most of the time, CPE is purchased by a consumer, often from a consumer electronics retailer, and is expected to work with CPE purchased by other consumers. Clearly, a purchaser of a fax machine can use it only with others who also have a fax machine. Likewise, a purchaser of one of the new picture phones can use it only with others who have the same type of device. Consumer expectations are in line with this simple state of affairs. In general, this works very well. Telephone service is a "common carrier" service because the phone system carries for a fee messages that are created by others. The telephone company has no involvement in those messages and carries them independently of their content or the sender or receiver. This carriage is to be nondiscriminatory. The telephone network is rarely "full." It is almost always available to additional users.

Cable service, on the other hand, consists of signals "broadcast" to a large number of receivers from a central point (headend) in a geographic area. There is very little—if any—communication up the cable system, except for some of the new two-way interactive services. Even in those situations, the signaling is asymmetric, with huge quantities of information flowing "downstream" from the headend to the subscribers' homes and comparatively little ordering or interacting information flowing "upstream" to the headend.

The cable service is very much like that of a publisher. The cable operator obtains programming, selects which to carry on its limited capacity, and makes content decisions based on feedback from subscribers. If the content decisions are well-made, the cable operator is rewarded with subscriptions. Poor content decisions result in fewer subscriptions. Just as a publisher has a limited page count to live within, so the cable operator has limited channel capacity. Once the cable system's spectrum is full, additional signals cannot be added without an expensive technological upgrade. There are limits to how much any upgrade can achieve. Spectrum is likely always to be a scarce resource. Many believe that the cable operator is a "First Amendment Speaker" and is protected in that role by the U.S. Constitution.

As an electronic publisher, it is critical that the cable operator have an audience which can receive the signals in a high-quality manner. The publisher who wishes to provide signals which require in-home hardware with capabilities beyond those incorporated in the equipment subscribers own, must overcome that deficiency in order to have customers. For example, if the cable system carries more channels than the subscriber's TV can tune, the cable operator must ensure the availability of a converter, or the upper channels go unwatched. Likewise, if the subscriber's TV has inadequate internal shielding and suffers from DPU interference, supplementary equipment must be provided to compensate for the deficiency. Simply put, the electronic publisher needs an audience in order to have a business! The audience requires proper equipment in order to participate.

Architectural Differences

It cannot be stressed too strongly that there are fundamental differences between telephone systems and cable systems in their basic architecture. Telephone systems are point-to-point connections with a switched path set up for the phone call and then taken down afterward. Two customers making repeated calls probably will never use the same path twice. The exclusivity of the connection between the two calling parties tends to make the communication more secure. In addition, if one of the consumers on a call behaves in a disruptive manner, it would almost never affect anyone beyond the other party to that call. Also, there is no danger of signal leakage from the plant or CPE which would cause hazards or inconvenience.

Cable systems, on the other hand, involve large numbers of subscribers sharing the same path and spectrum. Security very much depends on modifying the signal in a manner so that only those who are authorized can make usable results. Additionally, if one subscriber disrupts the network through the use of faulty equipment, he may impair the reception of many others or even cause a hazard for those who use the radio spectrum.

Even so-called switched cable architectures involve switching to clusters of 500 to 1,500 homes. There is still a large commonality of connection. Many subscribers have simultaneous access to the same signals. Cable "switching" is very different from that used in telephone systems.

The nature of cable architecture gives rise to its great economy and efficiency. Rather than using switched physical circuits, cable uses channel selectors to create "virtual circuits" from the headend to the subscriber. Much less wire and equipment is needed to accomplish this result, and much more of these facilities are shared rather than dedicated to individual subscribers.

Service Innovation

Cable systems have a long history of trying innovative new services—and failing. In the current method of doing business, the subscriber is able to participate in those experiments and vote with his dollars. If he likes the new service, he rewards it with continued subscription; otherwise it fails. The subscriber is protected against a loss of his investment in hardware—which is no longer needed when the service failed because the hardware is owned by the cable operator, not the subscriber. Even if the service concept succeeds, the service brand may fail. For example, there are at least five different versions of electronic program guides currently being offered. It is impossible to predict which will succeed and which will fail. But it is fairly certain that there won't be five of them five years from now! There may be two or even one, but not five. The consumer must be allowed to participate in the process of selecting the winner. But it is best if the consumer does not have to directly absorb the cost of finding the winner.

This kind of experimentation and innovation in services benefits the consumer and is an integral part of cable's history. Rapid and simultaneous deployment of in-home hardware is necessary for such experiments to take place. This is very different from the telephone model.

Consumer Expectation

The expectations consumers have about the in-home hardware they own must change if innovation in services and technology is not to be hampered. Traditionally, consumers have viewed consumer electronics equipment as having a long life—ten to fifteen years. Not only is the equipment expected to last that long, but it is expected to be useful as well. That is, the services which make the equipment useful are expected to survive that long. Cable operators are, on the other hand, accustomed to implementing service trials by providing the in-home hardware. If the service succeeds, more hardware is ordered and the service is rolled out. If it fails, the equipment is scrapped, but the subscriber is protected from having had to pay for equipment that has no further use. If future subscribers choose to purchase in-home cable hardware boxes, they must do that with the understanding that they have assumed the economic risk if the service is no longer offered.

Policy and Technology Requirements

In creating policy and technology requirements for the set-top box, there are three guiding principles which should be applied. The set-top box should:

- First, cause no harm; recognize that there are genuine hazards.

- Second, protect the intellectual property rights of those who make it all possible and worthwhile.

- Third, protect the privacy of consumers, particularly where return channels are employed on the cable system.

- Fourth, facilitate experimentation, innovation, and diversity in services, programming, and technology.

A few more words are in order on the issue of causing no harm. The set-top box should function without:

- Causing interference with other users of the radio spectrum such as aircraft communications and navigation, and emergency radio services (fire, police, rescue, etc.); i.e., these CPE products should not leak broadband cable signals from their internal circuitry.

- Causing interference with other users of cable service in other residences by feeding back internally generated interfering signals into the cable system.

- Causing interference with other users of cable service in the same residence by feeding back internally generated interfering signals into the cable system.

- Suffering picture degradation due to direct pick-up of signals via inadequately shielded internal circuits.

All these requirements and policy issues are complex and difficult to realize. But they are important if the information superhighway is to fulfill its promises.

4

Consumer Electronics

HOWARD MIROWITZ

The birth of the consumer electronics industry is linked in this chapter to broadcast radio and television. Although we think of hundreds of novel electronics products and games in this category, it is true that the most pervasive electronic appliances serve to render broadcast media. The linkages between broadcast radio and consumer electronics, and now the cable industry, are far stronger and more important than ever before. Changes affecting one of these industries can profoundly affect the others. This chapter explores the linkages of these industries throughout their history.

Consumer electronics products are increasingly thought of as commodity items that are produced in staggering volumes that can be manufactured only by huge, automated, high-technology "cookie cutters." While consumer devices are indeed high-volume products compared to many other markets, manufacturers contest the notion that all the products are the same. The notion of a quality range corresponding to price and features is described in this chapter and lends a new texture to the complexity of the next generation of new products: how can there be different quality levels with all-digital media? Clearly consumer electronics companies will figure out a way, just as they have with digital audio compact-disc players (although some say the range of quality is narrower than before, raising new marketing and design challenges).

The recent growth of home multimedia computers has prompted many to describe the conversion of the home PC from a business-based product into a commodity consumer product. This chapter reminds us that the transition of computers into the alien world of consumer appliances must follow a predictable pattern that obeys immutable market rules. These rules may indicate the future of the home computer, and that future probably will bear little or no resemblance to the device we use on the desktop today. In this case, the genes provided by the business computer ancestors are likely to be diluted and may even recede from view in the future as the more dominant consumer demands take over the design of future products.

Until the day this chapter was written, there have been identifiable media stream standards that, as a whole or in part, define the medium of exchange between the content creator and the consumer. Several key media streams are described in both the analog and digital domains. A new wrinkle is introduced with the addition of computer software code that controls rather than represents multimedia content. This new data type, if it can be called that, breaks the symmetry of traditional consumer electronics systems (except for the case of video games, which has its own set of closed rules). Standard transmission media and data representations used to define whole industry segments. The presence of incompatible control code might break the old model.

On the regulatory side, consumer electronics manufacturers are under less direct influence than cable and broadcasters, but are affected nonetheless. Choices and changes in media and transport mechanisms necessarily must be reflected in consumer products. But because of the commod-

ity-like aspects of the business, this industry must pay more attention to pricing policies to avoid antidumping or antitrust charges.

Finally, this chapter suggests that the future for consumer electronics manufacturers is confused at best. The shift from analog to all-digital, while it will take many years to effect, is full of uncertainty about consumer response to new products and services. The changes are larger and more complex than past incremental improvements to consumer products. The conversion from analog to digital may well be more disruptive than exciting to consumer electronics manufacturers because the delicate balances among costs, features, and customer acceptance are likely to be totally disturbed during the conversion.

History

A little over 100 years have passed since Hertz experimented with the waves radiating from electromagnetic fields, which Maxwell's theory predicted must exist. Five years later, Marconi called those waves "radiotelegraphy." Nothing in history compares with the century of technological achievements that has followed. Those Hertzian, or radio frequency (RF), waves are the basis for radio, television, magnetic and optical recording, telephone, and computer products, whose divergent histories are now beginning to intersect for the first time.

The consumer electronics era is most appropriately dated from the end of World War II, but consumers exerted influence on electronics markets long before this. The telephone was patented in 1876, and by 1891, when American Bell Telephone issued its first annual report to shareholders, the company reported a profit, a dividend, 132,692 subscribers, and exchanges in all but nine cities with populations over 10,000. The insatiable consumer demand for connectivity was restrained only by the speed with which homes and businesses could be wired.

Radio communications was the province of military and commercial interests until 1919, when an amateur operator in Pittsburgh began broadcasting music. The substantial mail response received from other amateurs came to the attention of a local department store, which began advertising the hours of transmission along with an offer to sell receivers.

Four years later, there were 500 music transmitters in the United States, reaching an audience of two million listeners. Reacting to the burgeoning consumer demand for broadcast entertainment, the U.S. government created the Federal Radio Commission in 1927 to allocate broadcast wavelengths.

Radio Corporation of America (RCA), which made radio sets as well as phonographs, diversified into content production and distribution by organizing independent radio stations into the National Broadcasting Company's Red and Blue networks. The Columbia Broadcasting System (CBS), on the other hand, was organized purely as a content company, taking advantage of the mass production and marketing of radios by RCA and its competitors.

The early pioneers in voice recording, communications, radio, and television technology are remembered today as much for their entrepreneurial skills as for their inventions. Bell, Edison, and Marconi all formed corporations to exploit their discoveries. They succeeded in attracting capital and management resources that enabled the creation of industry infrastructures, with the help of businessmen like David Sarnoff of RCA and William S. Paley of CBS.

Sarnoff, a Marconi apprentice who later became president of RCA, expressed possibly the earliest thoughts about a broadcasting infrastructure. In 1915, he proposed a large-scale music broadcasting entity that would build transmitters, create programming, and pay for itself through the sale of "radio music boxes." This idea was the seed—fertilized by RCA's investments—that grew into today's NBC and ABC networks. Paley, who founded CBS, went even further, realizing from his early experiences with radio advertising for his cigar company that broadcasting content could turn a profit on advertising sales alone. The fundamental technology for today's television broadcasting systems was developed in the CBS and RCA laboratories with funding from radio broadcasting profits.

Although television was demonstrated in France in 1909 and was used to broadcast the opening ceremonies of the 1936 Olympics, the first scheduled TV programming occurred during the 1939 World's Fair, where NBC and CBS broadcast three hours a week. By 1956, there were 450 stations and 37 million sets in the United States. The supply of content did not become an issue, for advertisers and Hollywood were more than willing to provide entertainment programming in addition to sporting events, news, and weather.

The relevance of this "ancient" history is that consumer electronics, broadcasting, and telephony have a mutual culture based on a long tradition. And despite all the technological and structural changes that have taken place in these industries over the years, they have maintained an unchanging de facto contract with consumers. A clear understanding of these points is essential in evaluating the extent to which technologies imported from other industries into consumer electronics might or might not lead to real business convergence with those industries.

Recent History

Traditionally centered on radio, audio, and television "appliances," the consumer electronics product mix has undergone significant changes over the past two decades. During this period, a steady procession of new products has successively restored the sizzle to retailing just as older products began to reach market saturation levels. Many of those products have expanded the realm of consumer electronics beyond home entertainment to include communications and information-processing devices.

While many consumer products have more complex counterparts for business and professional use, one of the most enduring traditions of the consumer market is simplicity. No one needs to be trained in how to turn on a radio or television set or how to operate a telephone.

Consumer electronics products are distinguished by their self-contained functionality and singular purpose. Their value is built in at the factory.

Consumer electronics products certainly become more complex internally as time goes on, but, outside, very few changes occur from year to year. Programming content is becoming more sophisticated, as are the methods of acquisition and authoring, yet consumer products continue in their narrow purposes of playing or recording single streams of data. While there are some exceptions, the mainstream technology focus of manufacturers generally continues to be on higher audio/video fidelity and cost reduction, rather than on adding more complex functionality that could confuse and alienate consumers.

Consumer products can be categorized into two broad classes: those that are connected to a communications network and those that are media-based. Media-based audio products are obviously the most volatile. In terms of sizzle, they have moved in 20 years from vinyl records to tape (reel-to-reel, 8-track, and cassette) to digital compact disc (CD), minidisc and digital compact cassette (DCC). Each successive product introduction brought considerable improvements in fidelity, simplicity, and value, with little resistance from consumers.

Videocassette recorder (VCR) decks emerged from an inauspicious beginning to penetrate nearly 80 percent of U.S. homes by 1990. The first major obstacle to videocassette was the initial refusal by the entertainment industry to release its properties for commercial sale on VCR media. In a few short years, however, videocassette sales and rentals have become major elements of Hollywood's marketing and financial strategies.

The second major obstacle was the confrontation over data stream standards, eventually won by JVC and others that championed the VHS format with its longer recording capacity. Similar confrontations occurred in the early history of audio cassettes, and in both cases the consumer mainstream made it clear to the industry that it would not buy into product-media incompatibility.

Incidentally, the three-quarter-inch U-Matic VCR format, developed by Sony in 1972 prior to launching its ill-fated Beta format for home consumption, is still widely used in commercial applications.

Network-connected products receive RF signals on broadcast and microwave radio frequencies and via cable. RF signals—VHF, UHF, AM, FM—are transmitted in frequency ranges, or broadcasting bands, defined by the Federal Communications Commission.

Community Antenna Television (CATV), or cable TV, systems began in the early 1950s in rural areas that could not receive RF signals directly. The first systems had a five-channel capacity. The ability to distribute programs by satellite was a key factor in the 1970s penetration of metropolitan markets by CATV. Satellite technology enabled cable systems to provide independent channels from distant cities, as well as "premium" entertainment networks.

With many CATV systems now having a capacity exceeding 100 channels and reaching a combined total of 40 million basic subscribers (out of 60 million homes passed), cable operators today are a powerful force in home entertainment. CATV's impact on CE products, however, has been minimal—"cable-ready" tuners have become capable of receiving the increased number of channels.

Technology innovations have taken radio and television products in two directions since the 1970s: very small personal units and very large home theater systems. Pocket-sized, high-fidelity radio, cassette, and compact disc players have been highly successful retail products, while very small video receivers have not.

With the U.S. market very nearly saturated, television set retailing has become a replacement business, with emphasis on larger picture tubes and higher resolution. Plagued by picture quality problems, projection TV never expanded beyond its initial high-end niche, but development of larger direct-view tubes—up to 40 inches (diagonal)—has created a viable large-screen home market. Other notable television developments in recent years are stereo sound and picture-in-picture—the ability to view two channels simultaneously.

The consumer communications products business emerged from the breakup of the Bell System in 1984, which enabled independent manufacture and sale of home telephones in most every conceivable size, shape, and color. The sizzle products in telephone retailing in the 1990s are cordless and cellular technologies.

Just as telephones were not consumer products when rented by the Bell System, CATV decoder boxes are not consumer products at present. While freeing themselves from wired connectivity to the telephone system, consumers are, at the same time, shifting in large numbers to cable television. Apparently, it does not matter to consumers how electronic products connect with content, as long as they work with minimal human effort.

The face of the consumer electronics industry is not as sharply chiseled as it once was—due in part to channel overlap with previously distinct products. The Electronics Industry Association (EIA) Consumer Electronics Group now incorporates home computers, facsimile devices, electronic typewriters, personal word processors, and home security devices in its industry analyses. These products all have specialized sales channels, yet may also be found in the same stores with radios, TV sets, stereos, and telephones.

The EIA does not track electronic musical instruments or electronic cameras, although these classes of products are regularly seen at the annual Consumer Electronics Show and in consumer electronics retail outlets. The case for product overlap generally lies in the products' suitability for retail distribution, which hinges on simplicity of design, volume manufacturing, factory or dealer service, simple straight-forward sales and ease of use without special training, single-purpose functionality, and price elasticity.

Some electronic musical instruments—and even some more complex consumer products such as home theater components—do not fit into the basic retail paradigm. These high-end products are designed to appeal to professional customers, hobbyists, or high-tech aficionados. Nevertheless, even the distribution infrastructures for these products share many of the characteristics of "pure" retail consumer electronics.

Growth of Global Enterprise

Much has been written about how Japanese manufacturers achieved dominance in consumer electronics. Many in the U.S. still believe the Japanese merely imitated United States technology, while using their labor cost advantage to force American companies out of the business. Not enough credit is given to technical innovations and Japanese willingness to forego short-term profits in favor of long-term market development.

A good case in point is the videotape recorder (VTR), invented by Ampex Corporation in California in 1956. In the early days of television, live broadcasts originating on the East Coast had to be repeated for West Coast audiences because of the time difference. Ampex and 3M Corporation pioneered the VTR equipment that enabled the first recorded news broadcasts at a cost of $75,000 per unit.

Sony and Matsushita independently began working on a miniaturized VTR in the early 1960s, as did Ampex. All three achieved a measure of success, but consumer versions in the mid-1960s proved too expensive at $1500, as well as being difficult to operate. Ampex abandoned consumer product development, but several Japanese firms continued to work on transistorized designs, improved head technology, and new recording methods. The first videocassette recorder designed for home use was introduced in 1975 by Sony.

RCA made a brief bid for VCR market share in the early 1980s, but abandoned its development effort in favor of optical videodisc players, a product line which it also subsequently terminated. In Japan, however, Pioneer Electronics continued to invest in videodisc in the belief that its higher fidelity would eventually find a market. Although laserdisc players failed to capture consumer interest in the 1980s, they are making a comeback in the 1990s, primarily to Pioneer's benefit.

Among the many other innovations by Japanese manufacturers are the first solid state radios, transistorized TV sets, portable stereo players, and hand-held video camcorders.

Since the labor cost advantage enjoyed by Japan in the postwar period has all but disappeared, Japanese manufacturers depend more than ever before on technical innovations to maintain their position. The labor advantage in Asia now belongs to China, Taiwan, Korea, Hong Kong, Thailand, Singapore, and even Indonesia and India. Many Japanese companies have moved assembly operations to one or more of these offshore locations.

Local consumer electronics industries have emerged in these developing economies to capture an increasing share of price-sensitive export markets. Other leading Japanese manufacturers appear to have shifted their focus to higher-value niche markets. The combined assault on both high and low ends of the U.S. market has driven all U.S.-based CE manufacturers (except Zenith) out of business or into mergers with foreign companies.

The United States market and its openness to Asian imports has clearly been the major factor in the growth of Asian-based global enterprises. European nations have been far more protective of their domestic manufacturers in the postwar period, while, in most cases, maintaining government ownership of communications systems.

However, this protectionism has allowed European manufacturers to avoid taking the severe cost-cutting and market repositioning measures that Asian competition has continuously implemented. As a consequence, the presence outside Europe of European CE manufacturers is being eroded, although acquisitions of American manufacturers (RCA and GE, for example) have allowed companies like Thomson and Philips to retain a foothold in the United States.

But we cannot overlook European technology innovations. In addition to being the cradle of the science behind the inventions, European contributions include the first television broadcast in 1909, development of audio cassette, CD, optical recording, and many other pioneering efforts that are less well known, perhaps, because of their use in government, as opposed to consumer, electronics.

Profile of the User

There are approximately 95 million households in the United States. Some 98 percent of them own a television set and a radio. About 94 percent have a media-based audio system (CD, phonograph, or tape player). These figures do not include radios and media-based audio systems in automobiles. Virtually the entire U.S. population consumes CE products.

When most people think about television, radio, or audio, they focus on program or media content preferences. The devices and networks that deliver the programming do not garner as much attention as the larger behavioral, sociological, and demographic controversies connected with content. However, there is a discernible pattern in consumer preferences related to the quality of CE products.

Market surveys indicate that reliability and quality are foremost in the minds of consumers when making product purchase decisions. Price is also important, but it is secondary to perceptions of quality. Certain aspects of quality become fairly obvious in direct comparisons of competing models on the sales floor, but consumers tend to rely more heavily on acquiring information about manufacturers' reputations from consumer publications and through word of mouth.

Thus, there is a wide range of tastes in appliance ownership, just as there is in content selection. Personal taste is exercised not only in the purchase of each type of appliance, but in the mix of appliances. The American family is ever less likely to be unified in its choices of entertainment or its schedule, resulting in secondary television sets and audio systems. Other common sights in today's society are joggers wearing earphones, teenagers carrying their boom boxes, and sports fans, perhaps addicted to instant replay, taking TV sets to a football game.

Although electronic appliances appear to be commodities, there are many niche markets that motivate manufacturers to innovate and to differentiate their products. Three types of self-motivated consumers—early adopters, hobbyists, and "prosumers"—represent niches that are likely to be key to the near future of digital convergence.

Early adopters are critical to the launch of any new technology. These are the people who bought television sets before there was a wide choice of continuous programming;

who bought VCRs during the VHS-Betamax wars; who bought CD players when stores were still filled with vinyl records and cassette tapes. They are the innovators who create demand by word of mouth, if the product is satisfying, and who quickly kill a product when it is not.

Hobbyists are hands-on people who get involved in the technology and achieve a measure of self-actualization from customizing electronic systems to suit their own tastes. They buy and integrate components from different manufacturers and enjoy collecting and trading information with each other. The first home computers were bought or built by hobbyists, as were the first stereos, CB radios, and electronic musical instruments.

Prosumers are very meticulous in their demand for the best of everything that has become an established trend, but—unlike hobbyists—they do not wish to get involved in the technical aspects. They derive self-esteem from owning products rather than from building or configuring them. They enjoy products that are judged by experts to be the best. They look to dealers or independent consultants for turnkey installation. This is the market for home theaters, and many prosumers also own home computers, fax machines, copiers, and modems.

The first generation of consumers for any new technology typically marvels at the technology itself and gains satisfaction from merely turning on a device and enjoying a new experience. The first consumers of radio and television were no doubt less critical of programming and more excited by receiving any kind of program, or even (in the case of radio) by composing and transmitting content themselves. This category of early adopter is often a hobbyist.

The second generation tends to be more responsive to the quality and quantity of content and becomes aware of product differentiation to the extent that it delivers more and better programming. The relationship between the consumer persona and CE products changed when high-fidelity sound, color picture tubes, many-channel receivers, and other innovations created primary entertainment sources that were psychologically comparable to a movie or a live concert or theater performance.

In the third generation, entertainment appliances influence, and become integrated into, the physical and emotional fabric of daily life, as the telephone, television set and radio are today. When architects, builders, furniture makers, and interior decorators are professionally influenced by the interaction of people and their entertainment appliances, or when neighbors gossip about the lives of soap opera characters as if they lived down the street, it can be concluded that the third generation has begun to live in the audio-visual medium. At this stage, consumers' emotional attachment to the products is enhanced by the availability of programming to match an ever-broadening spectrum of personal tastes, and the continued development of new products that further heighten the reality of the audio-visual experience.

There are many other niches and nuances in the mainstream consumer public that dictate product diversity—far too many to enumerate here. Most people will buy the highest quality product they can afford or the one that best suits a special need. These needs may vary locally or internationally. The TV-VCR combination is an example of a product that has sold very well in Japan and other places where living space is at a premium, but has not sold nearly as well in the United States, perhaps because it works against the U.S. consumer's desire for a diversity of choice.

Distribution Systems

The diversity of CE products and retailers often blurs a clear and concise description of distribution practices. Video products comprise the largest market segment, according to EIA, followed in order by audio products, home information, accessories, blank media, and home security. Home information, which includes personal computers, electronic typewriters, telephone products, and facsimile machines, is the newest and fastest growing segment.

Early retailers of television sets tended to departmentalize video with major (kitchen or laundry) appliances, resulting in the habit of calling TV an appliance. Audio systems were sold by separate departments or specialized dealers, who created a merchandising aura around fine distinctions in sound fidelity.

Over time, the merchandising of television sets as CE products has become separated from major appliances among retailers that sell both, even though such businesses may refer to themselves as major appliance or TV/appliance dealers. As kitchen appliances become increasingly controlled by electronic circuitry, distinguishing among appliances by calling television or stereo systems "electronic" and kitchen appliances "electric" has become somewhat atavistic.

The face of CE retailing has changed considerably over the past 20 years. National and regional entities now account for the bulk of sales at the expense of specialized independent dealers, the mainstay of 1970s distribution. The 200 largest CE retailers in 1991 accounted for more than $31 billion in U.S. sales.

The six dominant retailing channels, listed in order of market share, are:

- Mass merchants, such as K-Mart, Montgomery Ward, Wal-Mart, and Target.

- TV/appliance superstores, such as Circuit City and Silo.

- Electronics specialty stores, such as Radio Shack, The Good Guys!, The Wiz, and Adray's.

- Warehouse membership clubs, such as Sam's, Price Club, Costco, and Pace.

- Catalog-showrooms, such as Service Merchandise, Best Products, Consumers Distributing, and Sharper Image.

- Department stores, such as Sears, Federated, May Company, and Dayton Hudson.

Other types of businesses represented in the list of the 200 largest retailers include toy chains, drug store chains, electronics/photo and electronics/music combinations, military exchanges, software chains, and office superstores. The annual CE dollar volume of the top 200 ranges from $5 million to $2.7 billion, and the number of retail outlets per business entity ranges from one to 6,842.

Although declining in number, the traditional independent audio and video dealer remains an important channel for manufacturers. Independents cater to the hobbyist/enthusiast market—which requires a high level of personal service as well as technical expertise, including custom installation—and are key outlets for products such as home theater components.

In addition to their focus on particular demographic classes, TV/appliance dealers differentiate themselves through brand specialization. While a superstore will carry a selective representation of many brands, specialized niche dealers may carry the complete product lines of two or three manufacturers. Their brand knowledge and personalized service are a competitive advantage.

CE manufacturers distribute products through direct sales staff, manufacturers' representatives, and wholesale distributors. Most brand name television manufacturers sell direct to retailers, while audio products tend to be marketed by rep firms. Retailers typically acquire low-cost items, such as pocket radios and tape recorders, from wholesalers. Mail-order or catalog merchandising to credit card customers is a small but growing channel, as is television home shopping.

Video and high-end audio manufacturers continue to cultivate marketing partnerships with retailers through the "factory-authorized" designation. Authorized dealers may be selected on the basis of their floor space commitment, demonstration facilities, and staffing with knowledgeable sales representatives. In return, manufacturers may extend sales and technical training, exclusive territories, co-op advertising support, and attractive discount schedules.

Each channel is characterized by a different business structure with a rigidly disciplined set of management and sales tactics. Each channel presents a different personality, sometimes subtle but distinctive, nonetheless, to the consumer. Successful dealers in all channels are intent on adapting to changing consumer preferences and new product technology, but very rarely consider changing their essential business models or basic selling formulas.

The distribution of home computers has undergone a similar transition from independent, knowledge-based dealers to volume CE retailers. But the CE channels did not really add new pre-sale or post-sale support skills or knowledge comparable to traditional computer dealers as a result of this transition. Rather, the growth of home computer sales through these channels has been enabled by the development of channel-specific products, which are carefully adapted to the strengths and limitations of CE sales outlets. For example, CE dealers generally maintain a separation between platforms and content. With the rise of consumer software chains, users can shop for computer applications and games independently of hardware, just as they do for video and audio tapes and CDs. Of course, the increasing computer literacy of the buying public also plays an important role, since many people who are familiar with computers do not expect the same amount of help from a sales outlet as those who are relative neophytes.

In any event, products—including computers—for mass merchants, department stores, and other CE channels reflect adherence to channel doctrine: functional simplicity, built-in value, a product that can be sold by staff with minimal training, no after-sale problems.

However, home computer merchandise still has not achieved the level of functional simplicity of a TV set. Buyers are still unpleasantly surprised when they find that the CD-ROM games they bought do not run properly on their PCs for any of a number of all-too-common reasons. But these problems will be less frequent with new generations of "plug-and-play" computer and software architectures.

The trend toward convergence of CE and home computer channels is clear. The convergence is being achieved by modifying computer products to fit the capabilities of the channels and not vice versa. It will always be important to understand the limitations on the complexity of products and content that can realistically be merchandised in this fashion. Of particular concern is the tendency of dominant CE channels to increasingly eliminate or minimize value-added and after-sale services as competition and consumer price consciousness continue to force increases in productivity and efficiency.

High-volume channels are not likely to change their personalities or practices for a product. They will wait for the product to conform to their individual formulas. The traditional CE customers for advanced technology products are early adopters and hobbyists, and the traditional channel favored by these customers is specialty stores.

By contrast, the traditional customers for new computer technology are sophisticated end users who buy through value-added resellers, system integrators, or direct sales to internal value-added organizations, such as MIS departments.

Despite a history of precedents, it is difficult to assess retailer response to the multimedia revolution. Multimedia applications for home consumption are already on the market or announced, in advance of platform standards and (until recently) in advance of market demand. Speculation as to how multimedia product retailing will unfold is risky at a time when no one yet even knows what will be in the box that consumers carry home.

If history is any guide, early adopters will establish a new product/market category through specialty retailers who vie to be the first with anything new and relish the challenge of educating customers. At present, these early adopters are buying personal computer-based products. The market response to non-PC systems (e.g., 3DO and CD-I) is still in the process of developing. Philips' attempt to educate consumers about CD-I through home rentals of machines and content in video outlets, such as Blockbuster, serves as an example of the search for an appropriate specialty channel catering to early adopters.

Mass merchandisers will be attracted only by the existence of shrink-wrapped, volume-manufactured products in boxes that the consumer carries away to figure out unaided. Once mass merchandising is established, channel evolution may continue with the advent of a "category killer"—a super retailer who blankets the market with product and pricing that smother other channels.

Core Technologies

The physics of light and sound—which are fundamental to video and audio products—has been the basis of consumer electronics engineering concepts developed and refined over a 100-year period. If fully explored, this history could evoke the names and contributions of scores of luminaries from J. J. Thomson to Albert Einstein to researchers at Bell Laboratories.

The gift of the scientists to electronics engineers is the knowledge of how light and sound waves can be converted into electronic signals, transmitted over long distances or stored on a transportable medium, received or played back, and re-converted to their original form. In the traditional world of consumer electronics, the principal components in these transmutations are various types of transducers, modulators, filters, and amplifiers positioned in an analog signal path.

Transducers, modulators, filters, and amplifiers are core technologies for all consumer electronics products. Countless improvements have been made in all these areas over the past 30 years, including innovations enabled by LSI and VLSI circuits. Of course, this excludes the many engineering achievements that apply specifically to the creation and distribution of content.

Throughout an entire audio, radio, or TV system—including everything from the recording or broadcasting studio to the consumer product—the creation, storage, transmission, and reproduction of audio and video signals involves numerous processing stages. Although some products can both record and play or send signals, most consumer electronics appliances are either receivers or players, and, until recently, utilized analog, rather than digital, technologies for signal acquisition, processing, and storage.

Transducers were invented in the nineteenth century by experimenters like Marconi, who converted sound to and from radio frequency waves; Bell, who converted sound to and from electrical voltage levels; and Edison, who converted sound to and from mechanically created indentations on wax. Twentieth century inventors contributed magnetic and optical transducers.

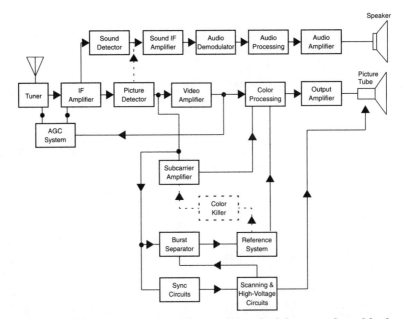

Figure 4.1 **Fundamental analog color television receiver block diagram. Courtesy of Mitsubishi Electronics America, Inc.**

An analog electronic system begins with a sensor that captures an audible or visible event. A microphone is a typical audio sensor; the image orthicon tubes or charge-coupled-device arrays used in TV cameras are examples of video sensors. The sensor acts as a transducer, converting the event to an electrical signal—a wave form that is always proportional to the size or power of the event. The output signal of the transducer is amplified, maintaining the power ratios of the original, which means that any spurious noise contaminating the signal is also amplified. This noise may be random, for example, caused by solar flares, or it may be introduced as an artifact of the system's own operation.

Analog systems include other circuits designed to filter out the unwanted noise and isolate the desired signal. Both signal and noise may occur at both high and low frequencies or at certain groups or ranges of frequencies. Thus, there are different kinds of filters (e.g., bandpass, notch, and comb) to create a clean wave. Still other modulators may shift the whole signal to a different power level. If the output is to a storage device, another chain of modulators, filters, and transducers is used to convert the processed electrical signal into physical changes in the storage medium, such as variations in the magnetic field on a tape.

To play stored signals, the process is reversed. For example, in an audio player, physical changes in the storage medium are converted to electrical signal variations again. When the electrical signals are sent to speakers, which are also transducers, they activate electromagnets, or other means of moving diaphragms, to make the compressions and rarefactions in the air that reproduce the original sound waves.

While an analog audio signal has one (mono) or two (stereo) wave forms, an analog color video signal is transmitted on a carrier wave that delivers three channels of program audio, a black/white luminance signal, a color subcarrier composite signal, synchronizing and timing information, and closed captioning data. The luminance signal indicates brightness, while the chrominance signal is a composite of hue and saturation quantities. The luminance signal alone corresponds to a black-and-white picture.

Using filters, the receiver must separate the audio and closed caption information from the video color signals and then separate the color signals back into the three primary colors, with values for hue, saturation, and brightness. In a normal TV set, the separated color signals are applied to three electron guns in a cathode ray tube (CRT), each of which controls a different primary color. The guns produce electron beams with intensities varying according to the fluctuations of the color signal voltages.

The inside surface of the CRT's viewing screen is coated with three separate sets (red, green, and blue) of phosphors. The phosphors are illuminated by the voltage-regulated electron beams through a mask, or aperture grille, that contains thousands of tiny holes or stripes. (The mask prevents leakage of energy from one point on the screen to another.)

Timing and control circuits in the receiver perform inverse calculations on the synchronization information. The electron gun "paints" the screen horizontally and vertically under the control of a set of electromagnets called a deflection yoke which steer the electron beam according to the outputs of the timing and control circuits. A complete picture is a scan of 525 horizontal lines (U.S. NTSC standard) repeated vertically 30 times per second.

In a CRT projection TV set, each primary color signal drives a separate CRT containing only phosphor that responds to that color. The three CRT images are projected on the TV screen, superimposed on each other.

Color CRT design strives to improve viewing image fidelity through resolution (sharpness), contrast, and brightness. The shadow mask and its refinements, especially the development of the aperture grille, are the principal developments in recent years. The mask's array of holes or stripes is created on a photo master by a laser plotter. It is applied to the faceplate by a photoetching technique. The number of holes or slots, which must always exceed the line resolution of broadcast signals, is contained in a suggested industry practice published by the EIA. Each manufacturer uses its own shadow mask design. Higher resolution is achieved by reducing aperture size, as well as the thickness of the mask, both of which are technically demanding processes.

Other contributions to picture quality have been gained through improving the lens characteristics of electron guns and deflection yokes, as well as their overall configuration, and in the preparation of phosphor material. Recently, liquid-crystal displays (LCD) have been used to replace CRTs in projection TV systems. They give a sharper picture, because the accurate registration of the three primary colors is built into the LCD panel and does not depend on projection angles. LCDs have also enabled small portable TV devices and have replaced the range-finder monitors in the latest camcorders. Research into other flat display technologies, such as plasma matrix devices, is directed at retaining the size advantages of LCDs while providing increased brightness for direct viewing.

A digital audio or video system requires an analog-to-digital (A–D) converter in the signal path leading from the input transducer to the signal transmitter or filters. The A–D converter samples the amplitude and/or frequency of the signal at very high speed and calculates binary numbers representing the strength or power of each sample. Thus, instead of an analog wave, the audio or video signal becomes a stream of binary-encoded data. A digital system further requires the inverse function—a digital-to-analog (D–A) convector—in the path leading from the signal receiver to its output transducer. Once the signal has been digitized, it can be stored or buffered in temporary memory, and all the required filtering and modulation can be performed by special-purpose microprocessors called digital signal processors (DSPs).

Digital technology has been successfully implemented in certain audio products, but digital video products present a far more complex set of issues. In the final analysis, most of these issues are related to cost and marketing, but the engineering challenges are noteworthy. First is the A–D sampling rate, which must be at least twice the frequency of the signal. Digital audio CDs, for example, contain adaptive delta pulse code modulation (ADPCM) samples at a frequency of 44.1 KHz. For the U.S. NTSC video broadcast standard, A–D sampling requires a converter operating at 35.MHz. For proposed digital HDTV systems, the converters must run much faster.

Next is the sampling accuracy, which is limited by the bit stream bandwidth. While 8 or 16 bit accuracy may be adequate for audio sampling, video sampling can entail 24, 32, or even 48 bit accuracy, with quality improvements at each level.

High bandwidths are necessitated by the volume of data generated by sampling at high temporal and spatial resolutions. A single color NTSC video frame, when encoded, represents about one megabyte of data, which translates into data traffic of 30 megabytes per second. Digital HDTV will require up to four times the bandwidth of digital NTSC. While data compression can be applied for transmission or storage purposes to reduce the required bandwidth, the actual image capture and display must be done with decompressed data. The compression and decompression is also typically performed by the DSP chip.

The cost of the consumer product depends upon the application of leading-edge manufacturing and design to systems with these performance levels. These advances, combined with future innovations in data compression/

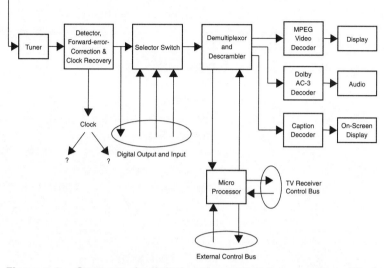

Figure 4.2 Conceptual digital color television receiver block diagram. Courtesy of Mitsubishi Electronics America, Inc. *Source:* EIA R-4.1 ATV Interface Subcommittee.

decompression methods, are expected to eventually make full-frame, full-motion, interactive video possible at prices consistent with consumer expectations. Cost reductions will depend on high-volume production, which is driven by a highly price-elastic consumer demand. At this stage, the question asked by CE manufacturers is not so much how to build a digital high definition television set, but rather who would buy it at a considerably higher cost than comparably-sized analog sets.

What is the benefit of high-definition TV to consumers? One answer advanced is obvious: the higher definition picture. HDTV is achieved by increasing the number of lines scanned to the screen from 525 to a proposed standard of 1024 to 1160. Increasing the resolution of the picture improves the image quality of large-screen TV sets more than it does for standard-sized sets, because picture imperfections are magnified on larger displays.

While analog HDTV technology, as seen in the Japanese MUSE system, allows significant improvement in resolution, digital technology expands the potential not only for higher scanning rates, but for other ways to enhance a picture. In addition, it unlocks the possibility of combining video with computer-generated information in new forms of content.

From the point of view of cable and broadcast network owners, another advantage of digital technology incorporating advanced compression techniques is the ability to provide more channels in the same transmission bandwidth. Network operators maximize their profits by offering a wider variety of specialized content designed to appeal to ever-more-finely segmented audiences through the willingness of advertisers to pay premiums to reach selectively targeted viewers. The networks may be reluctant to trade digital NTSC channels for higher quality HDTV channels at a 4:1 or 6:1 ratio, based on their economic analyses.

However, the ultimate authority is neither quality as specified by the engineer nor the variety of content available in the market, but rather perceptions of quality and value as seen by consumers. If consumers do not experience any improvement in the actual picture, and the additional channels do not add measurably to their viewing experience, they will not be willing to pay much more for digital TVs than they do for analog sets.

Other core technologies specific to VCR and audio products lie in the domain of magnetic and optical recording. Magnetic tape recording has a long history dating to the late nineteenth century and has far too many technology subsets for an adequate discussion here. As previously noted, transistorized components were the key to penetrating the consumer VCR market in the 1970s. New recording techniques were equally essential then and are even more essential to the new era of digital CE products.

One of those technical milestones was helical scan recording, which enabled reading and writing a complete image frame with each head pass. The previously dominant quadruplex scan method read and wrote segments of a frame at each pass. The continuing benefits of helical scan recording include increased recording capacity and the enabling of special effects.

Other milestones of the 1970s were the ability to record a color subcarrier signal on the same tape with images, and the ability to substantially lower the tape running speed of consumer models to accommodate feature-length movies in the compact VHS format, while maintaining acceptable picture quality. These examples illustrate the deep interplay of electronics, head design, electromechanical subsystems, and recording format standards required to achieve reliability and fidelity in analog VCRs.

In addition, improvements in media properties contributed to some of these advances. Higher coercivity media, such as chromium dioxide ("metal") tape, offered higher magnetic flux densities, without which the achievement of slower recording speeds would have been much more difficult. Most recently, digital audiotape (DAT) and digital compact cassette (DCC) have extended the capabilities of digital audio by allowing consumers to make their own digital recordings. Digital VCRs are already used in studios and are expected to appear as high-end consumer products within the next few years.

Optical recording technologies improve both the reliability and fidelity of CE products by eliminating the head-media contact and by increasing the precision of head-to-track alignment. These two characteristics also enable much higher recording densities than are possible with magnetic media.

The video laserdisc and compact disc or CD audio players are CE products using optical technology. Laser disk technology, originally developed in France, has evolved to high-volume data storage on 12-inch optical disk media, and, more recently, compact optical storage subsystems using smaller diameter media have emerged.

CD-ROM media is a current focal point for the convergence of computer and CE technology by virtue of its ability to store and play video, sound, text, graphics, and animation. However, while CD-ROMs are used in computing for general-purpose functions, addressing specialized market niches with single-function products appears to be the rule for CE product developers. CD-ROM media are used in electronic pianos, still photo reproduction, games, books, and graphics applications. Each of these applications has a different, single-purpose consumer product (or product family) dedicated to that specific function.

Applicable Standards

CE appliance manufacturers have traditionally been concerned with three types of standards: rules and regulations of the Federal Communications Commission, recommended practices published by industry and professional organizations, and de facto standards that are independently agreed to by manufacturers and program content suppliers.

Historically, these standards have applied only to the format of data (analog or digital) or its physical recording or transmission media. By contrast, standards in the computer industry apply to executable program code, as well as to data formats and media. This difference will have an enormous impact on the cultures of the two industries, and the degree to which they may have to change as their technology bases interpenetrate.

While radio and television frequency spectrum allocation is controlled in the U.S. by the Federal Communications Commission, the United States is one of the few countries in the world that allows the private sector to control broadcasting system technology and operations, and set most equipment operating standards. In the absence of statutory requirements, the industry has, generally, conformed to standards that are essential to a contented and growing market.

Industry recommended practices may be incorporated in FCC rules and regulations when deemed to be in the public interest. For example, the National Television Standards Committee (NTSC), established prior to World War II, created the broadcast signal format subsequently adopted by the FCC.

The development of color television prompted the industry to consider changing the format in the early 1950s, but a federal requirement for signals to be compatible with then existing monochrome sets forced broadcasters to innovate within the original standard. Following this pattern, the FCC imposed a similar downward compatibility requirement with NTSC on the new HDTV standard.

The 1953 NTSC color standard for the United States, later adopted by Japan, applies to chromaticity and signal coding. (Monochrome reception is achieved by setting R, G, and B signals to an equal value.) Two other standards must be considered in the international business environment. France and former members of the Soviet Union use the SECAM system, while the United Kingdom and other parts of Europe use the PAL system. The chromaticity of the three systems is essentially the same, with differences occurring primarily in signal coding. In all three systems, the display of the odd-numbered lines in the image alternates with the even-numbered lines in a process called "interlaced scan."

Efforts to establish a worldwide standard inevitably encounter the problem of backward compatibility with existing receivers, but export of programming by satellite and videotape is regularly achieved through signal conversion techniques.

FCC rules codify numerous appliance standards relating to "best performance in the public interest," such as manufacturer-supplied antennas, peak picture sensitivity, user interfaces for channel selection and tuning, and, most recently, the CATV cable connector and signal format. As FCC Class I devices, television, VCR, and videodisc units are also among those limited in both incidental radiation emissions and power output. Many CE products are subject to certification as complying with FCC rules.

Most legally enforceable CE standards merely establish boundaries that leave appliance manufacturers with a broad latitude for product quality differentiation. Recommended standards, published by industry associations, focus on technical issues and engineering practices. Prominent U.S. sponsors of standards committees that influence appliance design are the Electronics Industry Association (EIA), the Society of Motion Picture and Television Engineers (SMPTE), the International Electrical & Electronics Engineers Association (IEEE), and the National Association of Broadcasters (NAB). The conclusions of these private committees are often adopted by the American National Standards Institute (ANSI).

There are many noteworthy examples of de facto standards influencing CE appliance manufacturers. In videocassette technology, SMPTE, appropriately indifferent to consumer preferences, publishes standards for both beta and VHS formats. However, VHS became the de facto standard as the result of a wider selection of programming and independent manufacturer decisions to conform to VHS. The Philips audiocassette, developed in the 1960s, has been a de facto industry standard for tape recorder/player manufacturers since the 1970s. The Musical Instrument Digital Interface (MIDI), which governs connections and data interchange among keyboards, synthesizers, and other electronic musical instruments, began as an effort by companies such as Yamaha and Roland for professional applications.

Until the advent of compact disc players, all CE standards were based on analog technology. The Sony-Philips alliance on compact disc technology resulted in the first set of digital standards mutually accepted by appliance manufacturers and recording studios. Since that time, a new generation of digital audio standards, such as minidisc, digital audiotape, and digital compact cassette, have begun to vie for acceptance in the marketplace.

These recording standards cover a range of specifications, from the physical size of the package to speed to recording format. In addition, the original CD standard has been expanded in several directions to cover digital data other than audio signals, including various types of computer file formats. The International Standards Organization (ISO) maintains some of these standards as ISO 9660 (the "High Sierra" format for CD-ROM).

While more than 60 million homes are passed by cable broadcasting systems and over 40 million homes are actually attached, until the passage of the Cable Act of 1992 there were no enforceable national standards for CATV operations, although some local jurisdictions attempted to regulate their CATV companies—with limited effect. CATV rates are now regulated by the FCC, however, and the FCC has also announced that it will standardize the technology used for future digital transmission, as well as promulgating standards to ensure that existing analog CATV systems are compatible with consumer electronics products. The CATV analog signal conforms to NTSC standards for retransmitted programs, although programs originated by local cable operators may use any type of signal as long as locally provided converter boxes condition the signal for NTSC receivers.

The Cable Act of 1992, which enables consumers to purchase a converter box (among its many other provisions), has prompted a joint effort between the EIA and the National Cable Television Association (NCTA) to develop a set of technical standards for cable systems. Among the issues being addressed are channel allocation, type of signal, and decoder interface.

The latter issue addresses the problem of decoding conditional access programs by converters. As these devices are purchased by consumers, it is in both public and private interests that converters operate uniformly throughout the country, and that converters have well-defined interfaces with consumer electronics products.

HDTV and Digital Standards

High definition television is a concept originally advanced by the Japanese National Broadcasting Company (NHK) in cooperation with manufacturers of video equipment for professional use. The original NHK standard (MUSE) was an analog system. The successful application of HDTV to closed-circuit systems has generated worldwide interest in potential applications to consumer products. The fundamental visual differences in HDTV are:

- Vertical definition of more than 1,000 lines, compared to 525 for NTSC systems (625 for PAL and SECAM).

- A fourfold increase in the number of pixels, resulting in four times greater luminance.

- An aspect ratio of 16:9, versus 4:3 for NTSC, resulting in a 25 percent wider image.

- Improved color rendition.

- Closer viewing of the picture, which enhances the viewer's sense of participation in the programming.

- In some proposed systems, progressive (noninterlaced) scanning is utilized, as opposed to the traditional interlaced method, in order to allow greater compatibility with computer-generated video signals.

The U.S. mandate for HDTV broadcast signals by 2008 reflects a general confidence in technology advancements and a consensus of opinion that HDTV will be achieved by replacing the conventional NTSC analog signal with a digitally encoded signal. For all practical purposes, the MUSE standard has been eliminated from consideration in the U.S. in favor of a completely new system design.

The FCC is expected to adopt a set of digital HDTV signal standards as recommended by the FCC Advisory Committee for Advanced Television Systems (ACATS), much in the same manner as it codified recommended practices of the NTSC committee. There is no specific timetable for implementing ACATS proposals, but accepting the likelihood that systems will be deployed prior to 2008, FCC rules provide for simulcast transmission of analog and digital signals until that date. Several competing groups of companies initially submitted digital proposals to ACATS. Recently, the groups have moved toward a single set of proposals, referred to as the Grand Alliance.

Apart from HDTV efforts, a move is under way within both CATV and telephone industries to standardize a digital NTSC broadcast format. Several different methods are under consideration, including MPEG-1 and MPEG-2, developed by an ISO committee and advanced by both CATV and computer industry interests. While these methods allow digital video signals to be transmitted over the air or through a cable system, ADSL, developed at Bellcore and advanced by regional Bell operating companies, allows a digital video signal to be sent down twisted-pair telephone lines. The Digital Audio-Video Industry Council (DAVIC) is becoming the international coordinating body for much of the technical standards work connected with digital video-on-demand (VOD) systems.

SMPTE is the focal point for new video production and broadcast standards, which are better described as a completely new digital image architecture, including the data stream headers that will establish a standard for receiver compatibility. However, the data stream is just the tip of the proverbial iceberg. When video is perceived as a data type and the television set as a data processor, there is far more to be considered than just a one-for-one replacement of existing standards.

The Interactive Multimedia Association (IMA) model of the new digital world positions CE appliances as part of a continuum of applications bridging entertainment, information processing, and telecommunications technologies, with compatibility across all elements of the model. In this model, which largely reflects the thinking of influential computer hardware and software companies and content developers, CE products are assumed to be capable of executing program code that is loaded into them from communications networks or storage media.

CE products previously had only to conform to data format standards, leaving microprocessing functions completely within the purview of individual manufacturers. Because of this new interaction, CE manufacturers are being confronted with many classes of standards for the first time, such as:

- Interoperability with telephone systems

- Connectivity to data networks

- Compatibility with device-independent color encoding systems

- Compatibility with data storage devices

- Scripting formats to ensure compatibility of executable program code from thousands of different sources

- Multilayered bidirectional communications protocols

- Resident software standards (distributed object models, directory and file systems, operating systems, etc.)

- Complex user interfaces allowing control of multiple functions

Even standards familiar to appliance manufacturers are being debated in unfamiliar terms. For example, the familiar YUV/YIQ chrominance standard may not be adequate if consumers want to print their video images as well as watch them on TV. Digital audio standards for such products as compact disc, digital audiotape, magnetic minidiscs, and MIDI musical instruments also are already well established. However, new programmatic, electrical, and physical interface standards are required in any type of CE product architecture that integrates these digital audio functions.

Impact on CE Industry

As previously stated, the "technology continuum" concept is in sharp contrast to the traditional CE business model. While the consumer electronics industry fully acknowledges that it must rethink its position in response to the evolution of technology, consumer tastes, and new market opportunities, it does so with full awareness of responsibilities to fundamental economic interests.

CE manufacturers are active participants (and supporters) in the formulation of a new wave of HDTV and digital video standards. Clearly, standards related to interoperability— standards that allow digital consumer products to be purchased and used independently of the networks to which they are connected—are in consumers' best interests.

At issue are not standards *per se*, but rather the impact of standards on product design, manufacturing, pricing, distribution, and, most important, consumer response. Traditionally, standards have evolved from experience in product development and marketing. In the current situation, standards are being formulated in advance of any conclusive experience with consumer products and markets. The questions being raised by CE manufacturers are substantive, realistic, and in the public interest:

- How will compatibility with information processing and telecommunications systems affect pricing?

- What compelling new program content, derived from this compatibility, will motivate consumers to replace existing satisfactory products?

- Will compatibility result in uniformity and loss of value-added brand differentiation, as it has in the case of other digital products?

- How will retailers respond to paradigm changes, and who will pay for their retraining?

The larger portion of answers will have to come from CE manufacturers themselves. Yet, they cannot be addressed until a framework for decision making is established by standards committees comprising representatives from other industry sectors, as well as CE manufacturers.

Infrastructure

The infrastructure of the consumer electronics industry is more noted for its fragmentation than for a coherent model that simplifies technical and marketing decisions. A new wave of mergers and alliances—the results of which are not yet measurable—contributes to uncertainty.

The CE industry adds to its worldwide presence through interrelationships of multiple structures with overlapping interests. A summary of those structures follows.

CE Manufacturers

Asian companies dominate market share, but are in no way representative of a trade bloc. CE manufacturers are highly independent, competitive, and intent on product differentiation. They like to think that the fidelity with which a product records or plays its data stream acts to position that product on a quality/price continuum.

High-fidelity products command higher prices and sell in lower volumes; lower fidelity products have lower prices and higher volumes. In the traditional analog domain of consumer electronics, continuous design and manufacturing improvement translate directly into improved fidelity, or into lower cost for the same fidelity, and thus into improved product positioning. Companies can establish themselves at the low end of the market with a price-leading product, then move upscale with fidelity improvements, or they can enter at the high end, and expand their line downward with cost reductions.

Even as digital technology begins to pervade the CE infrastructure, the quality/price dynamics of analog products are still the foundation for the CE value-adding system, which has the following key elements:

- Each product manufacturer has freedom to unilaterally add its own quality, functionality, and other value to each class of product so long as the product meets relevant data stream standards.

- Each manufacturer can independently position products on value and price to appeal to different classes of consumers through distribution mechanisms appropriate to target market segments.

- Manufacturers support data stream standards that reinforce their freedom and independence, and that apportion value fairly between manufacturers and content suppliers.

- Manufacturers are keenly aware of consumer needs, wants, habits, and moods, and pace product introductions accordingly.

- Manufacturers carefully control incremental technology introductions to leverage existing distribution channels, because the cost of creating distribution exceeds the cost of developing new technology.

CE Distribution Channels

Mass distribution channels are optimized for high-volume sales of appliance-type products that are easy for consumers to install and use and require little or no support from the channel. Relatively small, specialized dealers focus on hobbyists, early adopters, and prosumers who require channel support in the form of advice, expertise, and configuration/installation services.

High-volume channels are more sensitive to price than any other factor, but adhere to the quality/price continuum by selecting the lowest-priced quality from competing manufacturers. Specialized dealers buy and sell in lower volume and are, therefore, more concerned with maintaining normal retail margins. They thrive on product differentiation and innovation that enable value-added services and advice to buyers to justify higher margins.

Content Providers

Entertainment content is provided to the home by several distinct sources and distribution channels with independent structures. The well-being of CE manufacturers depends on continuous improvements in content that influence consumer desires for higher fidelity appliances and for multiple appliances to satisfy a range of household interests.

Radio and television broadcasters may develop original content, buy from independent producers and/or join a network. The only requirement to be a broadcaster is an FCC license and studio equipment.

CATV operators, originally rebroadcasters, have more recently become content producers as well, in compliance with community access requirements, as well as high investors in companies like TBS, Liberty Media, Nickelodeon, and MTV. An NTSC broadcasting studio can be equipped for as little as $10,000, which has made possible numerous local interest channels. Major CATV operators have also invested in, or own, specialty networks designed specifically for cable viewers, in addition to a rapid increase in independent CATV-only networks. The impact of CATV on CE manufacturers is seen in tuners that accommodate an increasing number of channels.

More than 40 million viewers have demonstrated a willingness to pay a monthly fee to CATV operators to acquire a higher quality picture and/or a larger selection of program content than that offered by antenna reception. The advent of pay-per-view and its more flexible digital counterpart, video-on-demand (VOD), introduces another economic variable that could enhance the power of CATV operators as content distributors. It is a natural evolution for developers of nonbroadcast content to consider the CATV link to the home consumer.

Recorded content, such as VCR tapes and compact discs, is produced and distributed through a structure that traditionally has been far removed from CE manufacturers. The mutual dependency was recognized only in agreement on data stream formats. Those instances of entertainment content companies being acquired by companies with CE manufacturing units, such as Sony, Matsushita, and Philips, are seen by the CE industry as attempts to realize business synergy rather than as efforts to impose proprietary data stream formats on content.

While warehouse clubs and certain superstores sell both players and content, the two distribution structures for the most part remain distinct, and there is no indication that integration will become an industry-wide trend.

Telecommunications

The telephony industry is still another infrastructure that overlaps with consumer electronics—primarily through standards for home telephone, computer, and facsimile equipment. More recently, long distance carriers and regional Bell operating companies have advanced the concept of delivering entertainment content by telephone line, expressing the intent to replace traditional twisted-pair plant with fiber and coax lines.

In addition, cash-rich telephone companies have invested in, or merged with, CATV operators and content providers. A good portion of the movement toward "digital convergence" in the United States is being funded by telephone company capital. To some extent, this is occurring because the convergence-related business opportunities are less regulated than the telephone business itself, and, in addition, telephone companies may find it less expensive in many cases to buy a cable system than to upgrade a twisted-pair telephone system to the equivalent bandwidth capacity.

Information Processing

Both the computer and office products industries overlap CE distribution channels. Home consumption of electronic typewriters, word processors, and personal computers tends to fit the CE industry model of single-function products with the value built in. Computers, of course, introduce a different potential CE model, although it is uncertain to what extent that potential is realized by the typical consumer. Consumers who feel comfortable buying computers through a CE channel are much more knowledgeable about the computer's inner workings than they are about how a television or stereo product works.

In addition, people generally are willing to tolerate more hassle in connection with employment purchases than they will in connection with entertainment or recreation. Thus, to the extent that they use computers for working at home, they feel confident in buying a computer from a CE channel even though they will receive no more help or support than they would in purchasing a television set.

Retail PC models are often sold with preinitialized system software and bundled with application programs. The user loads different entertainment and educational applications in much the same manner as different tapes or CDs are loaded into their respective players. The trend toward cottage industries and increased work-at-home programs by major employers involves business applications by trained professionals, rather than consumers as defined by the CE industry. Market research surveys indicate that the chief purchase motivation of a home PC at retail is work (or education). Entertainment is clearly secondary.

The professional orientation of home PC purchases obscures any outright conclusion that the mass market for consumer products is becoming technically proficient in digital systems. However, the widespread use of computers on the job indicates that a large segment of white-collar (and an increasing number of blue-collar) workers are bringing computer proficiency home with them every evening. In fact, the same market research surveys also report that *after* purchase, home PCs are actually more heavily used for entertainment than for work!

Digital television broadcast (or narrowcast, as in VOD) could easily be accommodated by the existing CE industry infrastructure, despite the widespread changes in technology and equipment. Such an event certainly creates opportunities for new suppliers of equipment, components, and services to gain market share, assuming that digital television simply represents a new set of competitive dynamics and technologies unfolding in an old model.

The potential for structural change exists if that assumption is invalidated by the concept of programmable, multifunction appliances. In this multifaceted scenario the question arises as to who will provide the programming and other value-added functions, such as user interfaces for navigation and control—the counterpart of channel selection that will be required in the highly complex networks of the future.

Interactive television is already the basis for intrastructural changes and interstructural conflict. Digital systems open opportunities for automatic transactions related to merchandise sales, networked multiperson games, banking, and services such as travel arrangements, as well as communications applications (e.g., video telephony).

Content providers must install and manage sophisticated networks with an unprecedented number of nodes. Credit card verifications and "electronic money" impose another new set of network requirements. The brunt of handling such a huge volume of transactions falls to common carriers, who will have to increase switching system capacity.

A deployable large-scale network architecture to support fully interactive television is yet to be defined. Cable operators appear intent on a closed architecture, in which servers, software, and communications facilities will be under their control. In this solution, a successor to the converter box would house the navigation and control system, as well as the logic to interpret and render the content itself. The "set-top" box would be the consumer's link to the network architecture. Only the content (or application software) interface would be open.

Many consumer electronics manufacturers view the closed architecture as an infrastructure invasion. They ask the following key questions:

- What will external control technology do to the way television sets are designed and manufactured, and how will it affect the quality of the viewing experience?

- Will value added be perceived by the consumer as being in the control box, as opposed to the television set?

- What happens to the product line positioning strategy that is fundamental to retailing?

- Will the television business follow the path of the personal computer business, in which all products devolve into commodity manufacturing and pricing?

- Who will absorb the cost of retraining and repositioning distribution channels?

The regional telephone operating companies are historically accustomed to open systems due to having operated in a more heavily regulated environment which mandated support of independent customer premise equipment. The telephone companies might become potential open system architectural competitors to CATV operators. An open system would have several layers of hardware and software interface below the content level. Self-contained appliances, multifunction computers, or even servers owned by hobbyists or consumer aficionados could be attached to the network at all these layers.

To the extent that telephone company capital finances the development of communications networks incorporating digital convergence technology, open system architectures may become more prevalent. Yet, with multiple networks being deployed, nothing currently guarantees that the various "open" networks will be able to successfully interconnect at all their functional layers to form a seamless national infrastructure.

CE manufacturers have continuously demonstrated their advocacy of new technology that enhances the value of their products, the success of their channels, and the satisfaction of their customers. To achieve those ends amidst different visions of home electronics presented by other industries, each of which exercises a certain measure of real influence on the CE industry, represents an unprecedented set of challenges.

Convergence does not necessarily mean a total interlocking of current overlapping and abutting infrastructures. The traditional strengths and values of CE manufacturers may well be the harmonizing influence that enables digital systems to successfully penetrate the home market.

Regulation and Legal Constraints

Companies within the consumer electronics industry are subject to essentially the same regulations as any other employer, manufacturer, and marketer. In the only recent legislation targeted directly at appliance manufacturers, all television sets are required, beginning in 1993, to facilitate closed caption signals for the hearing impaired.

Because of their high volume of imports to the U.S., CE manufacturers are particularly attentive to Department of Commerce and Bureau of Customs regulations. National trade policies are subject to change in terms of both administrative enforcement and new legislation as the result of changes to the trade deficit, currency valuations, and lobbying by domestic industries that feel disadvantaged by imports.

CE importers come under regulatory scrutiny most often for their pricing practices. Setting transfer or wholesale prices too low can result in enforcement of antidumping regulations. Setting transfer prices too high can result in IRS claims of tax avoidance.

FCC regulation of broadcast signals indirectly forces CE manufacturers to comply with appropriate tuners. Until recently, the FCC regulated only RF broadcast signals. The Cable Act of 1992 put the FCC into the business of regulating CATV systems. The first round of administrative regulations resulting from the Cable Act were general, limited to covering existing analog systems. But a late-1993 announcement by the FCC indicates the agency's intent to "standardize the technology used by cable systems for digital transmission." The scope of FCC intent could range from mandating signal formats for compressed audio/video data to regulation of scripting languages for interactive software, although the

FCC initially intends to focus on data compression, encryption, and access control.

All other CE product characteristics are designed for voluntary compliance with industry standards, with the exception of television radioactivity emissions. Consumer concerns about potentially harmful television emissions prompted a regulation in the 1970s that requires CE manufacturers to keep a record of all sales for five years. Federal regulations also require each product to carry a label indicating where it was manufactured.

Underwriters' Laboratories (UL), a product safety testing organization originally established by the U.S. insurance industry, has a great deal of de facto influence on product design. Although UL is not a government agency, many customers, including the U.S. government, will not buy products that lack UL approval.

Regulations similar to FCC and UL exist outside the United States. For example, the Canadian Standards Association (CSA) fulfills these functions in Canada. In Europe, VDE certification is a requirement. In some countries, the imposition of these regulations, and the requisite bureaucratic procedures, act as nontariff barriers to imports.

Outlook

Consumer product manufacturers are approaching the most significant decision point in their history as the result of digital technology and applications that bring new and existing parallel infrastructures into competition and cooperation. Although CE manufacturers are noted for their willingness to take entrepreneurial risks, their key decisions of the 1990s are more related to institutional values than to the specific product developments that will follow.

The outlook is clouded by a great deal of uncertainty. The analog model of the CE industry will no doubt continue to be necessary for several years, resulting in a phased transition to the world of digital convergence. The timetable for change will be determined by a confluence of consumer, distribution, infrastructure, and technology trends.

The uncertainty of consumer response to digital products is the most critical concern of manufacturers. This is where cultural differences in marketing appliances and computer products are most evident. CE manufacturers have not historically announced products or promoted technology in advance of product availability. Both early adopters and the consumer mainstream demand instant gratification, or they lose interest. New products must work to perfection from the outset.

The computer industry, by contrast, is characterized by promotion of products that do not yet exist and by early delivery of products that do not meet all announced specifications. This "vaporware" phenomenon exists partially because companies seeking to establish technical standards related to executable code must engage in promotion prior to actual product shipment in order to attract the support of third parties, such as packaged software application developers and system integrators. While computer buyers expect and usually accept this behavior, home entertainment consumers are generally intolerant of undelivered promises. As we have seen, consumer electronics standards have never determined the executable code structures inside the products, but have been confined to data formats.

The consumer demand for instant gratification extends to product simplicity. The ability to "plug and play," based on an intuitive knowledge of the product, is essential for the mainstream customer, although the hobbyist minority no doubt enjoys a modest technical challenge.

The consumer persona is different for each CE product, just as it changes when a television watcher or music listener sits down to operate a personal computer. Will a superpersona emerge among users of multifunction products? Or will consumers continue to interact with products as single-function units, with the difference being a change of function on demand? In the latter case, every piece of content might offer its own user interface, and the consumer would attach only a nominal value to the multifunction system's top-level navigation and control interface. This subtlety would be of considerable significance to perceptions of value and product replacement decisions.

Meeting consumer expectations about cost is another area of uncertainty. Analog product engineering and manufacturing is a mature process no longer concerned with recovering initial capital investments. It is commonly believed that digital components cost less and simplify manufacturing, resulting in lower consumer prices. This is a valid long-term argument. However, analog products have already been through a long-term cost reduction driven by a cumulative production and engineering experience base of hundreds of millions of units. Digital systems will require additional components and completely new internal architectures, in which consumer product manufacturers may not necessarily realize the leverage of their prior analog experience.

In particular, digital systems will require manufacturers to develop or license operating systems, software, and firmware that are compatible with the content software as well as with the network software in future cable and telephone systems, or even the Internet. If the manufacturer licenses this code, its cost is essentially out of the manufacturer's control. If the manufacturer chooses to avoid the license royalties by developing the software internally, other issues arise. A good part of the cost of internally developed software is often incurred at the front end of product development, and increasing production volumes reduce its per-unit cost impact. However, CE manufacturers presently have very limited experience with managing software development, or with continuing maintenance and support for things such as field upgrades to later software releases. Considerable uncertainty, therefore, remains about the contribution that firmware and software costs will make to total consumer product cost.

The CE industry's analog business model is adaptable to digital products, in terms of consumer price expectations, if digital products yield higher fidelity and take their place on the same price/quality continuum as analog products. This condition implies a family of digital products, coexistent with analog products, all with appropriate levels of value-added differentiation.

Value-added differentiation lies at the crux of CE manufacturers' uncertain future outlooks. Companies such as Matsushita, Sony, Toshiba, and Philips have invested in content production and distribution, or in telecommunications or cable networks. Strictly speaking, these are really diversifications outside the CE business itself. An actual CE manufacturer does not produce or distribute content; does not develop software; does not operate networks; and does not rely on third parties to add value. But, because of the crucial role of executable code in digital content, these are all absolutely critical to the success of digital systems architectures in the marketplace.

What CE manufacturers do well is design, make, and sell hardware, which is differentiated through merchandising. The transition from analog to digital architectures will require manufacturers to rethink their definition of value added, or relegate themselves to manufacturing commodity analog input and output devices that would attach to the "real" digital CE system. In this scenario, consumers would most likely get the digital controller from other sources, such as a CATV operator, a telephone company, or a computer store.

Value-added, digital appliance manufacturers of the future must think in terms of executable code, random access storage, two-way communications, and network architectures. They do not necessarily have to deliver all these capabilities, but they do have to factor them into their business strategies, make the necessary alliances to implement those strategies, and try to capture as much of the value they add as is technically and economically feasible.

Value-added quality and data stream compatibility are traditionally separate issues. Product quality in consumer electronics has always been expressed in terms of the fidelity and reliability associated with brand names. Basic data stream compatibility has been an absolute requirement, not an arena for adding value.

In the digital business model, there are new types of media streams, delivered to consumers by new methods such as satellite broadcast, telephone line transmission, CD-ROM media, and, possibly, the Internet. These media streams will incorporate executable code in various formats (e.g., scripting languages), as well as media data streams themselves, also in various formats (e.g., MPEG2, ADPCM audio). The variety of code and data formats that establish an early foothold in the marketplace may create opportunities for CE manufacturers to add value by providing products compatible with a broader range of popular formats.

Data stream compatibility may be achieved through an external controller in some cases or by direct interface in others. There are two broad classes of interactive content: that which is delivered on storage media (locally interactive) and that which is delivered over a network (network interactive).

Development of the locally interactive class is currently seen in efforts to create an infrastructure for interactive CD-ROM products. Existing CE channels are suitable for content and player distribution, but content creation requires a commitment by authors to a specific platform and media stream format. At present, all such formats are proprietary to individual media and player manufacturers or operating system vendors. As long as authors are faced with the risk of backing platforms that may fail in the market, the development of an infrastructure for locally interactive products will be retarded.

The vaporware phenomenon, so noticeable in the computer industry, is migrating into the CE industry as those firms promoting proprietary CD-ROM media stream formats or system architectures seek to overcome this inertia. As we have seen, too much promotion in advance of delivering a wide assortment of tangible products that consumers can purchase and enjoy could alienate consumers and slow market development even when products become available.

An industry standard by which all CD-ROM media play on all manufacturers' products is essential to building market momentum for locally interactive entertainment. Such a standard might be achieved through the efforts of trade associations, such as the Interactive Multimedia Association (IMA), or may result de facto from the competitive marketplace, as has been the case in the PC industry with floppy disk file formats. Even with this constraint, locally interactive digital systems have the best chance of short-term growth in penetrating households.

Interactive digital networks have to deal with a far more complex set of interdependencies among infrastructures that are, as yet, incomplete or do not exist. A network plant represents a multibillion-dollar investment, and the payback appears at this time to require substantial revenues from unspecified services beyond home entertainment. CE manufacturers can only assume that an interactive digital network architecture will incorporate new layers of appliance control logic—yet more hardware and software. This additional control logic will reside partially in consumer products, partially in network-connected storage/computing/communications devices (such as video servers), and partially in the network plant itself in components such as ATM switches.

Exactly how network interfaced data streams will be presented to, and managed by, CE products depends on the shape of these future network infrastructures. The potential for multiple, competing content delivery networks—regional and national—with control logic and media stream format variations cannot be discounted. Achieving compatibility with all forces impinging on consumer products for interactive networked systems could be a monumental challenge for CE manufacturers.

The outlook focus thus turns to the standards formulation process. Digital and multimedia standards committees have an agenda that goes far beyond the concerns of CE manufacturers. We can hope that participation by CE manufacturers will contribute to compromises that satisfy the needs of large-scale data networks, business applications, consumer content, and consumer products. In the final analysis, CE product must, and will, conform to new standards. CE manufacturers have not yet defined a unified position on the impact of executable code embedded in content.

True interactive networks (as distinguished from interactively selected content such as video on demand) will increasingly rely on object-oriented technology, including scripting languages, such as Kaleida's ScriptX and General Magic's Telescript. In locally interactive content, all the software objects on which a piece of scripted content depends are locally present in the consumer product, either in its operating system or on the CD-ROM with the content. In interactive networks, on the other hand, a script might depend on objects distributed throughout the network. Different networks, or even different content servers on the same network, might support different scripting languages and different object models—methods for finding, using, and communicating with objects.

If CE products, then, are to be compatible with multiple networks, the products may have to be compatible with multiple object models and scripting languages. Thus, standards for object semantics—particularly in the areas of timing, synchronization, and fidelity of media streams—should be as important to CE manufacturers as they are to content authors and computer system/software developers.

Multimedia technology has created far more excitement to date in the computer industry than in the CE industry. In the computer industry, the ability to integrate video and audio as data types with text and image processing is seen as the source of an entirely new generation of business applications. However, CE products already process video and audio signals, and there is only a very thin, if any, rationale that text and still image processing capabilities will lead to increased sales.

It is even questionable whether digitizing existing audio and video content will really increase the absolute level of consumer demand—at least in the case of TV entertainment delivered over networks in the form of broadcast or VOD. The digital format does not necessarily increase video fidelity as perceived by the consumer. Digital television reduces the cost of delivering the content to the consumer only by allowing more content to be carried in the same transmission bandwidth, unless the network operator chooses to deliver fewer channels at HDTV resolution. In the early stages of the digital superhighway, at least, this is not likely to occur.

However, to an industry that has achieved market saturation with many of its products and is experiencing slow growth in others, any new product that ignites consumer interest is welcome news. The technology guidelines for gaining full cooperation from CE manufacturers are not overly demanding:

- Hide system complexity behind simple products.

- Decouple content from low-level hardware standards.

- Design fidelity ranges into digital software and content standards.

These technical goals, if achieved, will allow the merchandising of differentiated products to consumers and will allow consumer electronics companies to help ignite the digital revolution of the twenty-first century.

5

Handheld Devices

GARY BRINCK

Shifting from consumer electronics, this chapter addresses the so-called low end of the computing food chain: handheld devices. As we have seen in other industries, there is a great deal more diversity than one might imagine, and the resemblance to higher end computers is necessarily remote due to design restrictions and compromises necessary to make the products portable. Since computing devices seem at first to have been designed in a seamless continuum from small to large, handheld devices have, in this chapter, been appropriately defined as those that fit in your hand. This cleanly puts full-featured notebook computers on the "computer" side of the fence (and in Chapter 6) where the design and architecture have a clearer history and definition.

Handheld devices divide roughly into three categories: personal-assist, entertainment, and communications and information. Each of these categories has close relatives in other industries. Personal-assist products, such as dictionaries and

125

translators or dialers, serve the consumer electronics market and are therefore governed by its cost and distribution constraints. Entertainment products, including video games and miniature televisions, are the products with the richest multimedia features. Communications and information products, particularly those serving business markets, focus on more expensive services such as paging, telephone, and other information services.

The costs of miniaturization and consumer price expectations force handheld device designers to make frequent tradeoffs to optimize these devices into more or less single purpose products. Such optimization serves to advance the state of the art more significantly than the advances of any one product might suggest. So, to see the direction of the next generation of digital interactive multimedia products, it is important to look at the full range of handheld devices. The next wave of such products probably will aggregate the advances of the current crop.

Balancing the design parameters will remain an engineering challenge. Display size, resolution, weight, color, power efficiency, and available bandwidth form a very complex equation that will force innovation and optimization for years to come. These complexities are further complicated by the need for supporting infrastructures for communications capabilities. Once a clever communications transceiver is developed, an entire communications network must be constructed before it can be marketed. Therefore, these devices cannot be evaluated in a vacuum. As with broadcast, handheld devices that utilize radio spectrum fall under FCC control. As with the cable industry, privacy and data security are issues for many devices.

The author concludes with a compelling argument for the probability that handheld devices will not share commonality or compatibility with larger "desktop/laptop" computing devices. The basis for this prediction is not just that today's computer applications do not scale down well, but that handhelds are destined to become a totally separate class of media appliances with different utility and value.

What Is a Handheld Computer?

Scalability is an attribute of personal computer systems that most buyers simply assume. Most manufacturers of IBM-compatible PCs have learned (often quite painfully) that customers will accept nothing less than transparent scaling from subnotebook to Pentium, from low-end value leader to high-end server flagship models. This is no less true across the entire range of Apple's personal computers. However, at some point in any scale, the characteristics have changed so much that there is only superficial resemblance among the devices at the far ends and their cousins more toward the middle of the scale. Handheld computers are a case in point.

I define a handheld as a computer-based device that fits *conveniently* in hand, pocket, or purse and is *fully mobile* while in operation. It is the beginning of a new class of system that realizes the promise of the "anytime, anywhere" computing popularized by the computer trade press. No subnotebook, not even one with radio communications, meets this criterion because the scalability requirement demands that it have all the "PC" characteristics of its desktop and floor-standing kin. But at the next step down the size scale, user expectations suddenly change and those very same "PC" characteristics become irrelevant. This is the realm of the handheld.

Both manufacturers and users are just beginning to comprehend how the handheld transforms our notion of what a computer (and computing) is. In this chapter, we will briefly explore some of the characteristics of handheld technology, and how it is almost inherently multimedia.

History

Today's handhelds have evolved from several types of portable devices and tend to be application-specific rather than general computing devices. Attempts to scale the notebook and subnotebook computer down to handheld size have failed, largely because neither the display nor the keyboard required by their applications can adequately fit in the hand-held form factor. Handheld capability today is driven almost solely by packaging requirements for physical ease of use and transport. Whereas the size of the display and amount of storage in a typical desktop PC is dictated primarily by price, handheld devices are limited by the size and weight that can be conveniently accommodated by the user's hands, pock-ets, or purse.

Much of the origin of handheld devices lies in the con-sumer electronics world. Pocket electronics and portable entertainment devices have explored techniques and tech-nologies that are only now being visited by the general com-puter market with devices such as the Newton Message Pad, Hewlett-Packard 100LX, EO Communicator, and Zoomer. Examples include:

- Personal-Assist Electronics
  - Electronic dictionaries and translators
  - Franklin Computing's pocket devices
  - Electronic telephone directories and dialers
- Entertainment
  - Nintendo Game Boy
  - Sony Walkman and Discman
  - Microsize TVs
- Communications and Information
  - Pagers
  - Cellular phones
  - Two-way radios
  - Weather-alert devices

Multimedia in Handhelds

The size limitation of the handheld has long forced its cre-ators to explore differing types of media. Audio, motion (video on miniature TV-style displays), and touch-sensing input technologies are some examples. There is even a pocket-size printer technology. The AudioGuide Audiomate utilizes the audio technology from voice message systems, games, and desktop multimedia to produce a replacement for the portable AM radio, which previously was used to guide visitors around museums and parks. Visitors can now be given personalized itineraries made up of high-fidelity sound clips. Pagers have progressed from the simple beeper to units which incorporate LCD displays, vibrator (silent) alert, and even canned audio messages. Panasonic has introduced a pocket-sized checkbook calculator which even prints your check for you!

As with multimedia in general, the handheld arena is struggling to find the combination of function (application) and delivery (medium) that will cause buyers to get out their credit cards. These devices may not all be successful from a market viewpoint, but each is contributing to handheld multimedia technology.

Types of Handheld Devices

Business Data Collection and Distribution

Business data collection is a market that can economically justify handheld computer development. Increases in productivity, reduction in turnaround time, reduced error rates, and improved customer service are all possible with handheld devices. These are to be distinguished from notebook computers and presentation devices, which are also beginning to incorporate multimedia technology to assist businesses in marketing. The business data collection handheld tends to be a "blue collar" tool more than a professional one. Examples:

UPS DIAD Device UPS developed this substitute for the traditional truck driver's clipboard. Includes a stylus and separate display for capture of a customer's signature. A barcode scanner is incorporated to track individual package deliveries. UPS has recently concluded an agreement giving it a nationwide cellular phone network for communications.

Hertz Car Return This "belt-held" device provides walk-around calculation and printing capability, which enables Hertz representatives to check in a car and calculate and print a customer invoice.

IBM TouchMobile Developed for use in pickup and delivery application, the IBM TouchMobile handheld incorporates a touch-and-pen sensitive display with full graphics display and capture capability. A laser scans barcodes, and an infrared data link transfers data to other systems. The TouchMobile has no keys or switches of any kind—all interaction, even turning the device on, is via the touch-sensitive screen.

Laboratory and Plant Floor The need for measurement devices in a lab/industrial environment has led to a generation of handheld computers that can be carried to an instrument or tool, gather data or transfer new programming, and display at least basic information on the spot. Many companies are entering the market or developing advanced capabilities for devices which previously were little more than calculators with a memory.

Fujitsu Personal Systems (the PoquetPad), Grid, Intermec, Microwand, and Symbol Technologies are some others who develop devices for the business data collection market. Except for an occasional TV commercial by Federal Express or one of its competitors, there is little glamour or hype about these devices, but this is where the commercial bread and butter currently is.

Personal Assist Devices

The introduction of Apple's Newton Message Pad, the EO Communicator backed by AT&T, the Zoomer from Casio and Grid, and the Amstrad Penpad heralds a new era, that of the

Personal Digital Assistant (PDA). PDAs are intended to address the needs of the mobile professional by offering real-time message communications for people on the move plus a variety of personal services such as notes, calendars, and diaries. This is the consumer electronics market, with hundreds of millions of potential customers worldwide. None of the initial crop of these devices are perfect—poor handwriting recognition is one of the major complaints—but they have come close enough to the mark that more than 100,000 have been purchased by the "early adopters," who are willing to put up with some headaches in order to be the first to enjoy a new technology. The availability of these devices is also fueling a rapid expansion of mobile communications infrastructure and services.

Forecasts of future PDA sales range from 350,000 to several million yearly, depending on whether applications with broad appeal are developed. This search for the proverbial "killer application" is bound to include experimentation with multimedia technologies and some attempts to blend entertainment features with other, more businesslike PDA features. Some pundits believe that PDAs can be successful only by combining the several media types to which people are already accustomed—cellular telephones, fax, paging, and e-mail. Others conjecture on games, maps, FM radio, TV, and personal health monitoring as well. Each of these is already possible, and several can perhaps be combined in a handheld package. What combinations people will pay for remains to be seen.

Information and Communications

Portable CD-ROM-based devices such as the Sony Discman provide true multimedia capability, featuring video and audio with enough capacity for full-length titles. Exploiting the capacity of the CD-ROM in a carry-about package, the Discman enables voluminous information to be carried in the hand and viewed wherever the user may be—riding in a limo or working inside a jet engine! The lack of a broad selection of titles and the relative expense of the device has kept this one on the back burner, but industry acceptance of a standard audio/visual data format for CDs would make it more attractive. Still, the self-contained CD-ROM viewer is likely to be of interest only to those who for some reason cannot be near the ubiquitous TV screen—a worker in the field who needs reference information, for example. With today's low cost of mastering CDs, it is practical to publish technical and academic literature for a small audience, and multimedia techniques could enhance the information dramatically. The minisized CD makes it possible to design a truly handheld CD device, but it would be suitable only if a small viewing area is adequate.

Some companies are developing specialty devices which combine audio and visual aids in a handheld package. One example is the Garmin GPS 95, which offers video maps showing your current location relative to airports and air navigation beacons. It uses the Global Positioning Satellite (GPS) network and is intended for small-plane pilots. With a different database of reference points, the same device could easily show course and direction to any landmark or even underwater locations. Another specialty device is the Audioguide Audiomate, described previously. These are basically single application devices, preprogrammed with multimedia data.

Palmtop Computers

Palmtop computers are distinguished from personal-assist devices in that they attempt to offer more traditional computing and programming capability. Examples are the HP 100LX and the Poquet PC. Palmtops have not met with much success so far because notebook and desktop applications do not scale down to the palmtop display size and new styles of applications have not yet been perfected. The personal-assist class of handhelds with new applications and new human interface techniques now appears to be the direction of the market and palmtop PCs will probably converge with them in the near future. New, lighter weight subnotebooks will handle the role of the mobile, general-purpose computer.

Profile of the User

As with most computers, there are actually several classes of users with distinctly different needs and expectations. The "road warrior"—the traveling business/professional person—gets most of the attention in the trade press, probably because the journalists are themselves in that group. However, there is an equally large group of nonprofessional workers who spend every day in the field and whose needs have long been unsatisfied, as well as an immense, largely untapped market of people with everyday wants and needs. This is the so-called consumer market for electronics.

Business and Professional Users

The business and professional users represent an attractive market for handheld devices and many companies are attempting products aimed at this user. Users in this class are trying to stay in touch with their clients and the office while on the road. The focus, therefore, is on messages and communications. The needs can vary a great deal by individual style and the nature of the business, but usually include a selection of the list below (some of which are overlapping or mutually exclusive):

- Pagerlike message alerts
- Voice mail access
- Telephone (cellular or similar)
- E-mail

These users may also desire personal-assist applications such as notepads, phone lists, expense recording, and access to services such as bill paying, financial data, and world news.

The Blue-Collar Information Worker

The information needs of the "blue collar" worker have long been neglected. A substantial portion of many workers' day is spent assimilating or collecting information. This is especially true of workers who leave the factory or office to work at client premises, remote company sites, or in the field. Service workers who go to the site of the job (appliance or business machine repair, cable TV installation, meter readers, electricians and plumbers, etc.) are an excellent example of this type of user.

The applications for this group are:

- Access and display of where-to-go information from a database (client name/address, service needed, appointment times, etc.)

- Data collection (what was observed or measured, actions taken, client input, services invoiced and moneys collected, etc.)

- On the job assistance (display of service manuals with images with audio annotation and perhaps even motion video in some circumstances)

- Video data capture and voice communications for interactive expert consultation while on the job

Children of All Ages

Nearly all of us like games and entertainment but seldom think of ourselves as "users" of computers in this role. However, an entire generation of people have matured in an age when both home and arcade video games are a common mode of entertainment, and there is a now a large market for portable electronic games.

Similarly, handheld electronic books and movies appeal to this same group, offering the possibility of replacing the ubiquitous paperback book and cassette headphones with handheld devices that display text, images, sound, and eventually full motion video. The Nintendo Gameboy is the progenitor of this type of handheld, and both Nintendo and Sega are pushing well ahead of the PC industry in putting powerful multimedia capability in small packages. Products based on 64-bit graphics engines with high-quality digital sound can be expected to hit the stores by 1995, and will be in handheld form soon after.

Distribution

A variety of distribution channels serve the market for handhelds. Traditional computer stores have not been a major source for this type of device, but this is changing as products like Apple's Newton become available and the computer community embraces mobile radio communications. Consumer electronics stores have historically been a major channel for handheld devices, originating with calculators and multifunctioned digital watches and progressing into phone list/dialers, dictionaries and translators, cassette and CD players, video cameras, etc. These stores remain a primary source and are motivated to expand their role as the population embraces electronic devices of all types.

Industrial and laboratory supply houses have also been a source for handheld devices for some time, catering to the need for portable lab equipment and measuring devices, especially those intended for rugged environments. As computers became essential tools, this channel has expanded its role to include small devices which could be programmed like (or from) a personal computer. Never a high-volume channel, this source is being eclipsed by the mass market channels as handhelds gain broader appeal.

Office Supply Houses

Office supply houses used to be a source of pocket-sized devices, though possibly more as gifts for managers and professionals than as legitimate tools. However, as the office supply store burgeons into a mass market outlet for communications equipment (phones, fax, answering machines) and personal computers, they become an important channel for business-oriented handhelds. In the future, they may also sell the communications infrastructure that supports them in the business environment, e.g., local radio networks (RF

LANs). Look for large office supply stores to be a major supplier to the hundreds of thousands of small businesses and independent professionals who shop there.

Mobile Network Vendors

Vendors of mobile radio networks have been forced to supply handheld equipment to create a need for expanded radio services, i.e., building their own market demand. Other communications vendors (cellular, cable companies) are also moving in this direction, but the trend now is for joint ventures and underlying technology development rather than wholesale or retail sales channels. Look for this source to dwindle as a market channel, but become increasingly important as a market maker.

Core Technologies

Power Efficiency

Handheld equates to mobility, and power efficiency is the key. Efficiency in power consumption is necessary to reduce demands on the power source (batteries) and cooling. An efficient power source is required to minimize size and weight as well as to provide a practical period of use. Power considerations permeate the design of handhelds and cut across all technologies employed in them.

Low-Power Electronics and Peripherals

Low-power electronics have become commodities thanks to the popularity of notebook and subnotebook computers. The 5-volt technology which was with us for so many years has given way to 3.3 volts as a standard for digital components, and even lower voltage components are on the way. Lower power peripherals are also becoming available in the form of miniature hardfiles and PCMCIA cards.

Batteries

Battery technology was essentially stagnant for many years, with rechargeable nicad (nickel-cadmium) the mainstay for portable devices. The recently developed NMH (nickel-metal hydride) battery is now taking over because of its superior life and the environmental benefit of a minimal cadmium content. Other technologies, such as lithium ion, are beginning to reach the market and may offer further improvement.

Until recently, only rechargeable batteries were suitable for handhelds because it was not practical or economical to replace batteries every few hours. Reduced power consumption has changed this and handhelds using AA and camera-style batteries are beginning to appear. Replaceable batteries have the advantage of minimal downtime when power runs out, a key consideration when the handheld is a necessary tool for a worker. No charger to carry around and no waiting for a recharge.

A major trade-off in handheld design concerns battery weight versus operating life. With today's technology, the amount of power stored in a battery (usually measured in milliampere-hours) is essentially a factor of density and size. More power equals more battery and more weight. The practical limit for a handheld device is about a kilogram (2.2 pounds), with a weight under 0.5 kilogram (one pound) favored. Until there is some breakthrough in battery technology or an alternative power source is discovered, we will be plagued with handhelds that are either heavier than we like or limited in time of use. This power constraint will affect multimedia in handhelds—extra chips such as digital signal processors and heavy processing loads when manipulating large bitstreams all eat away at the power budget.

Power Management

Power management is a new technology which seeks to improve overall power efficiency by shutting off or slowing down components that are temporarily idle. This new technology was stimulated both by the battery-life shortcomings of portable devices and public interest in the "green" (environmentally friendly) PC. Essentially a software technique, power management also requires that hardware components incorporate the means to shut down and restart individual pieces at any time, and/or low speed modes of operation.

Display Size, Resolution, Weight, and Color

When we think of multimedia, we mostly think of sound and motion video. The display of a handheld is limited in physical size to about 2.5 × 6 inches so as to fit in the human hand. Even at that relatively small size, more than half the population would find the device containing it to be somewhat bulky and inconvenient to hold. The trend in desktop displays has been toward larger, high-resolution color displays. Now the display industry needs to focus on very small displays, light in weight and low in power. Color is still a few years away because of power requirements and cost, but the demand for displays for game machines and computerized appliances guarantees that we will see continual enhancements in small displays.

Nintendo and Sega are becoming the technology leaders here. Computer component vendors are beginning to notice that there is some outstanding technology in devices they once disdained as toys, and also that there is a large market there for their products. Some cross-pollinization, if not outright convergence, is likely.

Packaging

A handheld device must be rugged in a practical sense. As owners of notebook computers are discovering, any device that is carried around is subjected to bumps and falls, the hazards of outdoor weather, and even getting sat upon occasionally. The rather basic packaging technology of the desktop PC will not do for a handheld, and computer manufacturers entering the handheld arena had better beef up their mechanical engineering staffs. The packaging must be durable. easy to clean, resistant to fluids, and light.

The other key factor in packaging technology is ergonomics and style. Handheld devices must be convenient to use even though they are small, and innovative techniques are needed for the placement and types of all controls. Style, which has pretty much been neglected by every computer manufacturer except NEXTSTEP, is an important factor for the personal-assist class of handhelds to be used by businessmen. Like the cigarette lighter and the pen, style as well as function will be a selection factor.

Mobile Communications Networks

Mobile communications is a critical technology for handheld information devices and the capabilities, infrastructure, and future potential of the various networks must be thoroughly understood by strategists and architects who wish to take advantage of wireless information delivery to handheld and notebook devices.

Bandwidth is fairly low today, generally 4800 bits/second, but is expected to increase to the megabit range around the end of the decade. This will be sufficient for the transmission of the stream data typical of full multimedia applications. Cellular (CDPD), the various radio networks, and satellite all offer different speeds today and are progressing at different rates, but the table below illustrates the bandwidth likely to be available to a typical user or application for the rest of the decade.

1993	4.8	Kb
1994	9.6	Kb
1995	19.2	Kb (1994 in selected areas)
1997	56	Kb
1999	256	Kb
2000	1.5	Mb

Remember also that mobile communications are packet-based and shared. By its very nature, mobility means unpredictable network loads, as users move about, and varying transmission error rates. The quality of service on a mobile network is even more difficult to guarantee than on a wired network, making transmission of time-dependent data streams (video frames or audio samples) chancy at best. Extensive buffering will be required for this type of data.

The major types of mobile data networks are:

- RF WAN (Radio Frequency Wide Area Network)

- RF LAN (Radio Frequency Local Area Network)

- Satellite

- Cellular telephone (including CDPD)

- PCS (Personal Communication Services)

Each of these has advantages and disadvantages, and none today are actually able to deliver the "anytime, anywhere" communications visualized in computer industry literature. The major factors in selecting your type of mobile communications are:

- Coverage area and quality: make sure all the geography you will be in is covered and reception is adequate. If you need coverage while inside buildings, make sure your candidates provide it. Do not assume that coverage of a city extends to its suburbs and surrounding rural areas or lakesides.

- Cost of transmission and equipment: cost per bit delivered varies substantially, and there are many pricing plans and negotiable volume discounts. Connections to your hub may have substantial equipment costs and monthly subscription charges. The cost of the radio and modem for the mobile units can vary widely, depending on type and capability.

- Interchangeability: Do you require geographic roaming capability or connections to services available only via a particular network? Mobile network interfaces are not standardized like the nationwide telephone system— your RF modem will work only with a particular network, and only with software designed for it. This situation will likely change under public pressure, but for the next few years you will have only proprietary solutions to chose from.

RF WAN

Radio frequency wide-area networks (RF WANs) were originally developed for two-way voice communications, a market now being served by cellular telephone. The RF networks, however, have tremendous potential as vehicles for mobile data transmission. The traditional leaders in this technology were radio manufacturers Motorola and E. F. Johnson, but newcomer RAM Mobile Data is rapidly building a nationwide network and will have the advantage of all modern equipment when it is completed.

SMR (Specialized Mobile Radio) networks exist in most areas of the United States but few are linked together and they do not all use an identical radio frequency. Therefore, they do not provide a nationwide roaming capability, and it is crucial to be sure your intended area of operation is covered if you select an SMR network. Motorola's Coverage Plus service links together about 750 SMR networks (using landlines), and thus is the closest to being a nationwide SMR.

RACONET is a value-added network built on top of the Motorola and E. F. Johnson physical networks. Raconet makes it possible to bridge across separate radio networks and uses a sophisticated communications controller to interface to the network and permit voice and data to share the same frequency. By linking together multiple existing networks, Raconet is able to provide service throughout the United States. Roaming, however, can be a problem since the underlying radio nets operate on a variety of frequencies.

ARDIS, a joint venture of Motorola and IBM, combined the extensive private networks of the two companies to produce a single nationwide network with excellent coverage—reaching about 80% of the population of the United States and Canada. ARDIS has the advantage of uniformity across the country so that one RF modem can be used anywhere, and also uses a radio frequency that is able to penetrate within buildings.

RAM Mobile Data, a joint venture of RAM Broadcasting and BellSouth, offers the *MOBITEX* service nationwide but is not currently as broad in coverage as ARDIS. But RAM is investing heavily in expanding its network, and the ARDIS advantage in coverage is rapidly being eliminated.

MONET is a bold attempt by Motorola to integrate all existing radio networks by providing a switching service among them. A user served by ARDIS can, for example, send via Monet a message to a user served by Mobitex in another area.

Satellite

Satellite networks provide broad geographic coverage and may be either shared or private. Typically a satellite network operator purchases a portion of a communications satellite's capability, perhaps one entire transponder, and shares it among multiple customers for a fee. Contention in your area may be high or low, depending on how many customers are sharing the available bandwidth. There is also a time lag in satellite transmission while the message is broadcast up to the "bird," then beamed down to the network operator's switching site. The switching site then routes the message to its destination, either by landline or another satellite transmission.

Major players in satellite-based mobile networks are Qualcomm with its Omnitracs service, American Mobile Satellite Corp., and Rockwell International's new Mobile Communications Satellite Service. American Mobile Satellite holds the FCC license to provide voice, fax, and data service in all areas not presently covered by regional cellular services. New dual mode cellular phones will access the satellite network if out of range of local cellular service. This service was expected to be operational in late 1994.

Satellite interface equipment is still too large and too expensive for individual handheld computers, but it makes a good backbone net for cars, trucks, and boats. Data delivered via satellite can then be passed on to nearby handhelds. In a few years, however, new Low Earth Orbit Satellites (LEOS) will make handheld units practical. Irridium, an international consortium led by Motorola, plans to start worldwide mobile voice and data service by 1998 using a total of 66 satellites. The time frame is probably optimistic for such a major deployment.

RF LAN for Buildings and Campus

RF LANs are used to provide local coverage, usually inside a building or on a campus. Typically the RF LAN is actually a wired backbone plus wireless hubs providing short range RF transmission. This combines the low cost and high bandwidth of wired communications with the flexibility of wireless in the desired area. Since radio transmission is very short range, it also makes the terminal equipment (the RF modem) relatively low in cost and power requirements. Examples of today's RF LANs include Proxim's products, the ALPS RadioPort, and the NCR WaveLAN.

The owner of the LAN standardizes on a type (just as one picks Ethernet or Token Ring in wired LANs) and the modem for one LAN type will not work with another. Furthermore, the RF LAN interface card which works at your office will not allow you to receive messages while driving home or traveling on business (you need a wide-area network for that), so an easily pluggable form such as a PCM-CIA card is called for.

The RF LAN is likely to be the first to offer enough bandwidth to make practical the transmission of audio and, eventually, video. University campuses have both the need and the capability and will probably lead in multimedia on the RF LAN.

Cellular Telephone

The cellular telephone system is the most widely established wireless communications medium, but in its present form is expensive and low in bandwidth. Also, the signal quality in many areas is inadequate for data transmission. CDPD (Cellular Digital Packet Data) is intended to remedy that, but CDPD requires enhancement of the thousands of transmitter/receiver cells across the country, and thus is still some time and many investment dollars away from universal availability. However, McCaw Cellular, a major cellular service supplier with 104 markets in the U.S., has announced it will complete the CDPD upgrade in all its sites by mid-1994.

PCS

PCS (Personal Communication Services) is a new network service which should be superior to cellular for digital data transmission. Like cellular, PCS uses a lattice of radio cells to provide coverage of an area and the modem transmits to the nearest available cell. But PCS cells are much smaller than cellular telephone cells (telephone cells sometimes cover a 25-mile diameter), and therefore the radio in each cell costs less while offering better quality in its small area. Of course, it takes many more cells to blanket an area, so there is still a considerable investment in infrastructure. The FCC has set up the frequency band for PCS and plans are under way, but PCS does not yet exist as an actual network.

Network Services

In an attempt to establish an immediate market, some network providers are offering built-in services. One example is the Embarc Network from Motorola, which distributes news and weather from *USA Today* and industry news from Individual, Inc. on a more-or-less real-time basis to subscribers. It also has an e-mail capability that can link to your office. RadioMail, Inc. offers similar capabilities via its NewsFactory service on the RAM Mobile Network.

The current Embarc device is an attachment for a portable computer which performs radio communications and handles text only. Future Embarc interfaces could be integrated in a PDA and are envisioned to offer images and color, but network bandwidth as well as the cost of large transmissions may make this prohibitive.

Applicable Standards

Data Formats for Information Distribution

As with many information technologies, standards are lacking for data formats, making exchange of data difficult if not impossible. Handhelds, which rely heavily on data obtained from larger systems, are particularly affected by this lack.

Operating Systems and APIs

The current generation of handhelds are all basically proprietary when it comes to system software, APIs, and development tools. This is to be expected—most are new products attempting to bring new technology to market. The developers all promise an open system (and they must, to foster applications for their products), but nothing has yet been accepted as a de facto standard. The PDA class of handheld is all pen-based and each uses one of the popular pen-based OS. Game systems and specialty devices are pretty much unique, making it difficult to capitalize on shared software.

Mobile Network Interoperability

Incompatibility is rampant among mobile networks at all levels—physical transport, low-level and high-level protocols, application interfaces, everything. Standardization is a necessity to obtain the any-to-any capability that is envisioned, but, at present, each network is trying to establish itself as the standard and dominate the others. Motorola, which profits from the manufacture of communications equipment as well as selling network services, seems most willing to lead efforts towards interoperability. Frankly, the entire mobile and handheld arena exploded much more quickly than even its promoters expected, and the world is unprepared.

Data Transfer Port

Another standardization need is widespread acceptance of a very short range (wired or wireless) computer-to-computer transfer port so that handhelds could universally transfer bulk data to each other or to larger systems. Rather than sending everything via e-mail style messages, it would be advantageous to transfer quantities of data as complete files over something like an infrared light path with a range of 2–4 inches. It should be possible to set the handheld with its port facing another's similar port and transfer many kilobytes of data quickly once both systems authorized it.

Infrastructure

The primary infrastructure requirement to enable the widespread use of handhelds is the mobile network. As discussed earlier in this chapter, much of that infrastructure is in place or planned, but interoperability problems will slow progress.

Another sort of infrastructure involves the availability of applications that make the devices worthwhile. I foresee a need in two areas:

1 Applications for the PDA (Personal Digital Assistant) class of handhelds will struggle for some time, trying to find the functions that make sense and the style of human interface that makes them usable in a small package. This class of device will be mostly sizzle but no steak until this application infrastructure develops.

2 For an enterprise and for individual users, few things are more frustrating than lack of commonality across platforms. Existing applications do not scale down to handhelds, even if the operating systems and APIs were compatible. Imagine using your favorite PC application on a 2 × 4 inch screen with no QWERTY keyboard. Notebook-size systems are about as small as that style of interface can be used. Since many users will have notebook and/or desktop PCs as well as the handheld variety, it is inevitable that many people will

want the same services on all of them. Since the big ones will not scale down, it seems the only alternative is for the handheld variety to be offered in scaled up versions. This will happen once an application is accepted as a handheld standard, but will take some time.

Regulation and Legal Constraints

There are several issues concerning mobile networks that will impede the creation of a wireless "information highway."

- FCC Bandwidth Allocation: when, how much, and to whom? Advanced functions and greater bandwidth will require a greater portion of the frequency spectrum, but the process of assigning bandwidth is both political and technically complex. And the government will now sell access to bandwidth rather than simply give it to qualified candidates, finally recouping for the taxpayer some of the value of a public asset.

- One of the possibilities is a kind of giant RF LAN that could be possible if community networks were franchised similar to cable TV and cellular telephone. In fact, cable companies would love to get into the data network business and have the infrastructure to do so, if regulatory hurdles can be jumped.

- Privacy and data theft will be a major concern and further legislation to deal with this in mobile networks is sure to be required. Theft of data can be difficult to prove unless a copy can be found, even though events and actions may suggest strongly that someone possessed knowledge intended to be private.

- Data encryption will be necessary to protect the privacy of digital data on the airwaves. Existing private key schemes as well as future public key technology will permit this without noticeable overhead. However, the U.S. government's insistence that all future encryption be based on the "Clipper" chip is a burning issue that has all businesses and a growing proportion of the public up in arms. Clipper includes a "back door" which will permit an authorized federal agency to fully decrypt the data—the equivalent of a wiretap for encrypted data. The possibility for abuse makes nearly everyone reluctant to commit to place all future data under Clipper management. And the ready availability of alternatives to Clipper technology makes it certain that any organized criminals or foreign agents will be able to avoid government monitoring of their own systems, so only the public is likely to be exposed.

Industry Outlook

Handheld computing devices are on the verge of a major market explosion. The necessary technologies have evolved to the point where the devices are practical and the industry is struggling with the applications that make them worthwhile. Commercial applications (inventory, stock picking, package tracking) will lead the way because the applications are obvious and cost-justifiable. At the other extreme, entertainment devices will also proliferate as Sega and Nintendo introduce 32-bit graphics and improved sound. The personal-assist class of devices (PDAs) is likely to take the longest to mature, despite a high level of promotion by manufacturers and hype in the trade press. This is because viable applications are most difficult to define here and the subnotebook computer will always be a competitor, offering broader capability at a not-much-greater price. Other factors:

- The current generation of potential customers accepts and desires electronic tools to make their jobs easier or their life-styles richer. Usage of all types of handheld and portable devices will explode as prices drop well below $1000. PDAs are already priced in that range, but volumes will have to be huge for them to be profitable at that price.

- Miniaturization of components is adding to handheld capability on a regular basis.

- Notebook and "Green PC" technology is yielding mass-produced low-power components. At the same time, battery technology is improving, albeit slowly.

- Mobile network coverage and capability are expanding rapidly. Look for megabit data rates by the end of the decade. This will open up the possibility of video on demand.

The single largest inhibitor to multimedia in handhelds is the restricted screen size. Expect attempts to circumvent this constraint with techniques such as folding, multiwindow displays (useful for books and reference materials), and curved screens which can be programmed to give a panorama effect and the perception of greater size. Neither of these techniques is a general solution, and you should not expect them to be successful beyond limited niches. The electronic pocket book remains an impractical device, even though full color, sound, and motion video can be played and titles can be delivered via the minisize CD ROM. The industry has been hoping for a breakthrough in display technology to lick the size (and weight) problem and consumer interest in PDAs and pocket devices assures continued research, but it is impossible to predict what direction the solution will take or when it might come.

Future handheld devices will be divided into two general types—those that rely on communications and those that do not. On the self-contained (noncommunicating) side, the capacity, convenient size, and removability of the minisize CD ROM will stimulate industrial and academic information applications to assist field workers from geologists to appliance repairworkers. Many of these will also have communications capability, especially telephone, but the primary function will be information retrieval. Maps, blueprints, specifications, and such will be invaluable help, even though limited by small display sizes. Another successful niche will be the intelligent audio guidebook, which capitalizes on the thriving tourist and recreation market by providing talking guidebooks that can respond to requests for additional detail, skip unwanted topics, and show maps and illustrations where appropriate. These also appeal to park managers, who see an opportunity to reduce the need for presentations and pamphlets. A standardized interface for such devices would be a plus for the industry, if not the traveler—attractions could sell or rent guidebook CDs to travelers who use the handheld they purchased or rented for their vacation travels. The cassette tape attempted this concept and proved inadequate, but travelers continue to spend large sums on books and slides so there is no reason to believe the approach will not work if the medium is up to the task.

Communicating handhelds will exploit both wired and wireless connectivity. One likely development is an enhanced relationship between the ubiquitous telephone answering machine and personal assist devices. Intelligence in the home or office answering machine will filter calls, automatically forwarding selected calls to the handheld and providing responses to others customized to the caller. The handheld will provide pager-like messages and many will include cellular or other mobile telephone and both audio and data message storage. The lowly answering machine will become a personal voice communications management system and the PDA a mobile extension of it. Look for telephony to play a large part in any successful personal-assist device.

Do not confuse handheld computers with their ever-so-slightly larger brethren, the subnotebook, which is a true IBM PC-compatible personal computer. Future handhelds will digress from the PC genre and become a unique class of device, but with strong relationships to desktop and laptop PCs. Think of them as media devices and not as miniature desktop PCs and you will begin to perceive their value and potential.

6

Computer Hardware

MONTY MCGRAW

Pick up the paper today and read about "interactive multimedia," and there is a good chance the subject matter refers to multimedia delivered on a CD-ROM (Compact Disc Read Only Memory) and running on a multimedia-capable personal computer. Though there are clearly many flavors of multimedia, this category has received the bulk of attention, and for good reason: it is among the highest growth segments in the computer industry.

Multimedia computers have become consumer electronics-like commodities sold through new retail channels with pricing and promotions more appropriate to mass merchandising than anyone in the computer industry has experienced before. Now subject to severe price pressure and the risks of high-volume production, short product life cycles, and high inventory risks, manufacturers are having to learn to quickly adapt to an alien style of business. Meanwhile, from a design and architecture point of view, designers are irretrievably shackled to the business/productivity market that spawned the multimedia PC.

So where did multimedia computing come from? This chapter explores the details of the three major technology waves that have shaped today's multimedia computer. The first wave aggregated computers into similar architectures utilizing similar or compatible microcomputer chips; the second brought consumer-friendly features such as color and the ability to connect to television sets. The third wave emanated directly from the business community and established fixed architectures that have remained in force a decade later. Current multimedia products are still constrained by these evolutionary waves.

The dual application of similar technology in business and consumer markets is particularly apparent in multimedia computing due to the genesis of the technology. This duality has resulted in a great deal of confusion and frustration on the part of business-oriented companies, who are less adept at dealing with fickle consumers and their apparently unreasonable expectations regarding costs and features. While it would be preferable in many ways to throw out the old designs and start over, the capabilities and architectural limitations of multimedia computers cannot be easily changed or abandoned due to the need to remain compatible with the thousands of applications and software titles that have so far fueled market growth. Also, as was noted in Chapter 4 (Consumer Electronics), the introduction of hardware-specific computer program code as part of the multimedia data mix seriously complicates consumer marketing.

This chapter reviews the various types of digital audio and video as well as the compression schemes needed to fit these media types on current storage and transmission media. These data types and standards were established without government involvement, unlike many of the other industries. Through this market-driven process, one can sense an almost arbitrary establishment of standards that seems almost like design by Brownian motion. Formats and standards were established either because it was the best that could technically be accomplished, or, simply because one data type out sold all the others. No matter how the elements that comprise *a multimedia computer were established, they are now irrevocably part of the structure of interactive multimedia computing and must be complied with.*

In addition to focusing on multimedia data types, this chapter introduces the notion of "publishing formats," a concept new to the computing world. Computer architectures have been designed first and foremost to provide productivity tools. That means that computers are tools that users figure out how to apply to their own business. Users put words into word processors, data into data bases, and numbers into spreadsheets. When they are finished shaping and grinding these data, they print them out. The addition of multimedia support in these same computers inaugurated a shift to using the computer as a player of prepublished information. The establishment of more-or-less standard publishing formats is an indication of the fundamental change in the use of computers as a result of the addition of digital multimedia.

Finally, the incredible growth of home computers has created a new impression for where interactive multimedia are likely to be centered. More than a quarter of U.S. households now have home computers, most of which are multimedia capable, and that number is fast growing. The old debate about whether new interactive home services will occur on your television set or on the computer in the den may well be over. Home computers are now in the home, and they are interactive—televisions are not.

History

Introduction

Interactive multimedia has emerged as a direct result of the personal computer revolution. Seminal work in the 1970s by computer visionaries at the Xerox PARC research center foretold a future where computers would be more personal (everyone could own one) and more useful (delivering prodigious quantities of information and cross-references on demand) such Alan Kay's Dynabook and Ted Nelson's Xanadu concepts.

The development of microprocessors incorporating the essential elements of a computer on a single integrated circuit in the early 1970s enabled first simple digital calculators and digital watches followed by a flood of labor-saving devices like electronic typewriters and microwave ovens. However, these devices embed the microprocessor and are not designed to allow access to the microprocessor. Personal computers are designed to encourage access to the microprocessor and, therefore, open the immense spectrum of computing to individuals. Virtually any task accomplished on larger mini and mainframe computers can be tackled by personal computers, only limited by the quantity of disk space they possess. In fact, the performance of today's high-end personal computers has surpassed the performance of larger computers costing over a million dollars.

The basic microprocessor in a personal computer can operate only on information in digital form (binary bits with values of either 1 or 0). These digital data can be stored and retrieved from a variety of computer storage medium, including magnetic tape, magnetic disk (both floppy and hard disk), and optical disk (CD-ROM and magneto-optical). The data can be transported using removable media such as floppy disk or CD-ROM, and it can be transported over digital networks such as Ethernet, telephone lines, or cable TV lines, or broadcast such as digital cellular, infrared, or other digital radio techniques.

In this chapter, we will cover a brief history of the personal computer and then discuss digital multimedia and issues with respect to the architecture of personal multimedia computers.

Brief History of Personal Computers

In the mid-1970s the Altair represented one of the first personal computers, but the standard user interface was front-panel toggle switches and LEDs. It included the S-100 expansion bus, which provided opportunities for third-party adapter boards including TTY and video interfaces, but the system and similar clones attracted more industrial and business interest than consumer. One of the Altair clones was the

Processor Technology SOL computer, which contained a standard character video adapter and keyboard with a couple of horizontal S-100 expansion slots. These computers all used the Intel 8080 microprocessor. Standard sound capabilities were nonexistent or limited to bells or beeps and there were no standard graphics capabilities. Standard mass storage was cassette tape, and the eight-inch floppy drive emerged. These machines evolved in the late 1970s to CP/M operating system, but were still limited to less than 64 KBytes of memory space. Lack of graphics and audio prevented the first wave of personal computers from even rudimentary multimedia capability.

A second wave of personal computers, with plastic cases and standard keyboards and graphics, emerged late in the 1970s targeting consumers directly, including the Apple II, Tandy TRS-80, Commodore 64, and TI-99/4. Microprocessors included the MOS Technology 6502 (Apple II and Commodore 64), Zilog Z-80 (TRS-80), and TI 9900. All but the TRS-80 drove an external monitor or TV (TRS-80 integrated a monitor) and used audiocassette interfaces for storage. External floppy drives (5 1/4" 140 KBytes capacity singlesided) were eventually available for most of these computers. The Commodore 64 and TI-99/4 had game cartridge interfaces for ROM-based software, which were faster and cheaper than the extra cost of a floppy drive. Memory space was around 64 KBytes, usually split about equally between system ROM and DRAM.

Multimedia capabilities were primitive due to limited (if any) colors and small memory space. The Apple II provided some small 8-bit expansion slots. Microsoft delivered the BASIC interpreters for the Apple II and TRS-80 machines and began writing applications such as Adventure for the Apple II (and also developed some early memory expansion boards and processor enhancements). Although BASIC language was standard for these second wave personal computers as a user programming language, the bulk of the programs sold used assembly language for maximum performance and minimum memory space. This second wave of personal computers emerged as the pioneer multimedia machines, primarily for entertainment and some educational applications.

The third wave of personal computers was initiated in 1981 with the IBM PC. This machine contained a 4.77-MHz Intel 8088 microprocessor (1-MByte address space and 8-bit expansion bus) with 32 KBytes of ROM (including Microsoft Cassette Basic), 16 KBytes of DRAM (expandable to 64 KBytes on the system board), and a cassette interface. IBM boards for the 8-bit PC bus included video interface options such as the CGA (Color Graphics Adapter) with TV monitor output and RGB monitor output, 320 × 200 4-color graphics and 80/40-column text modes or the high-resolution Monochrome board with 80 column high-resolution only text. A single-sided 160 KByte capacity 5 1/4" floppy option was available with an 8-bit controller board. Serial and parallel interfaces were available on separate boards.

Microsoft DOS 1.0 was the operating system on a floppy, and memory expansion boards became successful very early as the application programs quickly outgrew the 64-KByte memory capacity of the system board. The Compaq Portable was the first PC-compatible computer, and its 256-KByte memory capacity, dual floppies, and standard CGA plus Monochrome automatically switching display board in a portable package was very popular. An IBM XT computer added a standard 10-MByte hard disk and double-sided 360 KByte 5 1/4" floppy drive to the original IBM PC.

In 1984 the IBM AT computer was introduced with a 6-MHz Intel 286 microprocessor and up to 256 KBytes of memory expansion (later stacked DRAM chips were offered for up to 512 KBytes memory expansion). A 30-MByte hard drive and 1.2-MByte double-sided floppy drive was standard. The 16-bit expansion bus (named the ISA bus by Compaq) supported the full 16-MByte memory space of the 286 and provided complete backward compatibility with the previous 8-bit bus of the IBM PC. The IBM EGA (Enhanced Graphics Adapter board) offered 320 × 200 4-color graphics, 640 × 350 monochrome graphics and high-resolution text modes—graphics memory expansion from 16 KBytes to 64 KBytes offered 640 × 350 4-colors and 320 × 200 16-colors. Compaq introduced the first 32-bit Intel 386 16-MHz personal computer and maintained full PC compatibility in September 1986 with a standard base memory of 1 MByte and expansion to 4 MBytes and the first EGA compatible video adapter. Bundled with the Compaq Deskpro 386 was the new Microsoft Windows 386 version 2.1.

Apple's Macintosh computer in 1985 offered a 16-bit 7.8 MHz Motorola 68000 processor with 128 KBytes of DRAM, 64 KBytes of system ROM, built-in monitor with high-resolution monochrome graphics (512 × 342), mouse, and a 400-KByte 3.5" floppy drive. Sound hardware included an 8-bit monaural digital audio circuit driven by screen refresh hardware. The Macintosh Plus increased the base memory to 1 MByte with 128 KBytes of ROM, a double-sided 800-KByte 3.5" floppy, and a SCSI port. In early 1987 Apple introduced the Mac II computer with a 32-bit 15.7-MHz 68020, 256 KBytes of ROM, 1 MByte of DRAM (expandable to 8 MBytes), a 32-bit NuBus expansion system, 640 × 480 16-color graphics (expandable to 512 KBytes of graphics memory for 256 colors), and stereo digital audio circuitry. Commodore introduced the Amiga computer with the Motorola 68000 processor, 512 KBytes DRAM, TV graphics output (320 × 200 4096-colors), 3.5" floppy drive and a mouse. This computer featured hardware graphics acceleration and stereo 8-bit digital audio circuitry.

A month after the Mac II announcement, IBM introduced the PS/2 series of 286 and 386 computers including standard VGA (Video Graphics Adapter 640 × 480 16 colors, and 320 × 200 256 colors), and 3.5" 720-KByte floppies. Compaq added a 100% VGA-compatible graphics adapter in late 1987. IBM also introduced the 8514 high-resolution (1024 × 768) adapter, which was incompatible with the VGA. One significant change in the PS/2 machines was the introduction of the Microchannel expansion bus (both 16-bit and 32-bit versions with up to 20 MBytes/sec performance) incompatible with the previous ISA bus adapter cards. Compaq led a consortium of PC-compatible companies to introduce the EISA (Extended ISA) bus in 1989, which provided full ISA adapter compatibility and new 32-bit 33 MBytes/sec high performance for new EISA adapters. Newer local bus interfaces like the VESA VL-Bus and Intel PCI bus offer even faster alternatives to the ISA bus for interfaces like high-speed graphics frame buffers.

IBM's leadership during the first decade of PC compatibles solidified the use of personal computers as a tool for business, and high price tags kept these machines from being used for entertainment or education except as a secondary use. The penetration of the PC compatible computers into the home during that first decade was primarily driven by business utility. However, PC price wars led by Compaq in 1992 have driven PC prices below $1000, and this is dramatically increasing the percentage of PC compatibles purchased directly for home use. Current trends for PC compatibles purchased for home use indicate increasing penetration of digital multimedia capabilities in the form of sound boards, Super VGA graphics boards, and CD-ROM drives. These multimedia hardware trends in addition to the emergence of low-cost multimedia software on CD-ROM and a wide variety of multimedia software development tools have enabled this third wave of personal computers to emerge as the leader in delivering interactive digital multimedia today, as shown in Figure 6.1.

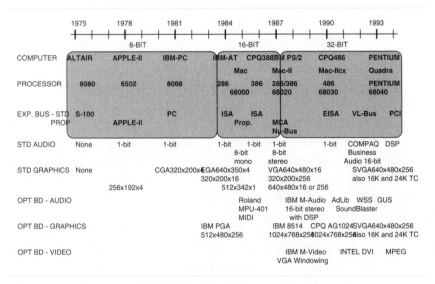

Figure 6.1   Multimedia hardware timeline

Multimedia: What/Who/Where?

What Is Multimedia?

Multimedia for our purposes delivers multiple types of sensory information, with aural and visual being the two most practical in today's technology. A movie or television show is, therefore, multimedia. Use of a personal computer provides the ability to interact and control the flow of multimedia information, unlike the linear flow of information in a movie or television show. This interactivity is achieved using a fairly simple input device to the personal computer such as a keyboard, mouse, joystick, trackball, or touch screen. The interactive multimedia user is able to control the flow of information only so far as the programmer allows through a special computer program called the script.

Early interactive multimedia efforts in the 1980s used personal computers to control analog sources of multimedia information such as laserdisc players. The Interactive Multimedia Association (IMA) created a standard specification to promote using laserdisc players controlled by personal computers to address such needs as interactive training and education. This standard has not penetrated the consumer sector due to high equipment prices and sparse development tools. Digital multimedia has several advantages over analog multimedia. One advantage is that both the multimedia information and the interactive script are in digital form and can be delivered on the same medium such as floppy disk, CD-ROM, or computer networks. These digital storage mediums are much more cost effective than analog storage, and digital multimedia data are more extensible into the future digital networks such as the information highway and new cable TV and telephone networks, which will deliver digital multimedia directly into the home.

There are currently a number of incompatible systems that deliver interactive digital multimedia to consumers:

Philips Compact Disc Interactive (CD-I) contains a CD-ROM and a 68000 microprocessor and connects directly to a TV set for around $500. The target market is consumers of both educational and entertainment titles (edutainment). The consumer is not allowed to access the microprocessor, and Philips stresses that this is not a personal computer. Philips has announced a cartridge upgrade for the CD-I player to view new MPEG digital movies on videoCDs by the end of this year.

Tandy Video Information System (VIS) contains a CD-ROM and a 286 microprocessor and connects directly to a TV set for around $400, also targeting consumer edutainment. The consumer also cannot access the microprocessor in the VIS. The VIS runs Microsoft Modular Windows operating system, which is similar to Windows but not completely compatible.

MPC is defined by the MPC Marketing Council as a PC compatible with at least a 386 microprocessor, CD-ROM drive, digital audio, and Super VGA graphics running Microsoft Windows. A computer monitor is required with these machines. These machines are available as complete systems or add-on upgrade packages, from around $1200 and up for the systems and $400 and up for the upgrade packages, which usually include a CD-ROM drive and controller and a digital sound board. The target market is consumer edutainment, plus, with full PC compatible functionality, these machines are complete personal computers.

Kodak Photo-CD contains a CD-ROM drive and microprocessor system connecting directly to the TV set for under $400. The current Photo-CD cannot quite be considered multimedia by our definition, but Kodak has announced plans to add digital audio capabilities in the future. The target market is consumer and business digital photography. A customer can have a roll of 35mm film developed and at the same time receive a special Kodak Photo-CD with the pictures stored digitally in various sizes. The Photo-CD player allows the customer to view the pictures on a TV, or the Photo-CD can be used at the film development store to make custom greeting cards and the like.

Kodak also offers software programs for PC-Compatibles or Macintosh computers with CD-ROM drives to view or edit these pictures and use them in presentations. The Philips CD-I player includes the ability to view Photo-CD pictures.

More products containing CD-ROM drives are expected in the coming months due to game-machine manufacturers desiring the economies of low CD-ROM production costs and the advantages of over 600 MBytes of digital information on a single disc. These CD-equipped game machines will still not have full personal computer capability with applications beyond games, such as word processing and calculation.

Who Uses Multimedia?

Special Effects, Design, and Visualization

High-end business applications of multimedia include development of special effects for movies and television and multimedia in product design. Recent movies such as *Terminator II* and *Jurassic Park* have featured special effects designed completely by using computers. Visualization of complex scientific phenomena has been enhanced by creating movies using multimedia techniques. New-product design is accelerated by using multimedia data created through computers rather than the more labor intensive building of models.

Presentation Graphics, Voice Mail, Computer-Based Training

Mainstream business applications of multimedia include enhancing presentations with multimedia audio and video, voice-annotated electronic mail, and computer-based training enhanced by multimedia. Advertising kiosks can be greatly enhanced with multimedia video and audio presentations. Computer-supported collaboration is also enhanced through sharing multimedia data.

Consumer Entertainment and Education

An emerging market for multimedia is in the personal entertainment and education industries. Recent examples of multimedia entertainment available for PC compatibles with MPC functionality include titles such as Virgin Games 7th Guest and ICOM Sherlock Holmes. Multimedia education titles such as encyclopedias (Compton, Grolier, Microsoft) and targeted education titles (Dinosaurs, Beethoven) are appearing on the market for PC compatibles with MPC functionality.

Where Can You Get Multimedia?

Retail and Mail-Order Sales

Multimedia platforms and titles are currently available through retail and mail-order distribution. Completely configured multimedia personal computers and optional upgrades for digital audio, CD-ROM, and video are both available. Current multimedia entertainment titles are available on floppy and CD-ROM. Some titles are exactly the same on both, though media with the CD-ROM version offer advantages in minimal installation time and minimal use of the personal computer hard disk. For some titles, like Sierra's King's Quest V and VI, the CD-ROM version enhances the game by adding digitized voices instead of just text dialog boxes and enhanced animations. Some of the latest multimedia titles are available only on CD-ROM, taking full advantage of the enormous storage capacity (over 600 MBytes per CD) to deliver a rich digital interactive multimedia experience. Multimedia authoring packages are also available for personal computers through retail and mail order.

Workstation Customers and Applications
(High-End Applications)

Some software developers are taking advantage of the emerging availability of CD-ROM drives for workstations to offer titles on CD-ROM. Primarily these are photographic clip-art images for professional publishing or multiple software packages for limited trial use before purchase. Very little software is available for workstations specifically targeting multimedia authoring or playback.

Multimedia—How?

Platform Requirements for Digital Multimedia

Digital multimedia platforms come in a variety of forms, which can broadly be divided into playback and authoring categories. Multimedia authoring platforms are more expensive than playback platforms because of additional hardware for audio and video capture. Authoring platforms also tend to require more processing power and memory and this also increases their cost. Multimedia platforms can be completely integrated units such as MPC or CD-I or can be optional packages which upgrade nonmultimedia personal computers to multimedia playback or authoring capabilities.

Multimedia implies multiple types of data such as video, text, audio, and graphics. A multimedia application orchestrates these data types typically under interactive control of the personal computer user. The size of digital multimedia data and their real-time stream-oriented nature have implications for the architecture of a multimedia personal computer system. Digitizing video and audio is the process of capturing and converting typically analog data (broadcast TV, VCR tape, broadcast radio, or cassette tape) to digital information.

Personal computers can process data only in digital form. Examples of the need for the personal computer to process the data are compression/decompression and editing. The vast size of uncompressed multimedia data (almost 100 GBytes for one hour of 640 × 480 true color uncompressed images at 30 fps plus stereo CD-quality digital audio!) essentially mandates compression of both the video and audio data streams in order to deliver this type of content on a cost effective delivery medium such as CD-ROM. Digital networks will require higher compression levels than CD-ROM in order to allow the network to be shared with multiple users.

Digital Audio Data and Compression

Audio data, which include voice and music, consume from 8 KBytes per second (telephone quality voice) to 176 KBytes per second (stereo 16-bit CD quality at 44.1 KHz sampling rate). These data rates can be multiplied by sixty to calculate the digital data-storage requirements for one minute of audio data (from 480 KBytes to 10.5 MBytes per minute for telephone versus CD-quality data respectively). Audio data can be sampled, companded, or linear, with varying accuracy. Digital telephone circuits operate with 8-bit accuracy on companded (logarithmic) data. This sampling is also found on some workstations and personal computers.

The most popular audio data circuitry for personal computers is 8-bit linear sampling at 11 KHz or 22 KHz (Mac and Sound- Blaster and compatibles). Very high-quality music is recorded for Compact Disc players with 16-bit stereo linear sampling at 44.1 KHz and some PC audio boards now provide this capability. In general, increasing the number of bits per sample increases the quality of the recorded data, and increasing the sampling rate increases the frequency response of the data.

Digital data sampling yields a frequency response of no more than one-half of the sample rate, which is one of the reasons CD-audio data are sampled at 44.1 KHz in order to deliver a frequency response of up to 22 KHz. This frequency response is a bit better than that of the human ear and is the new standard for high-fidelity sound set by digital audio CD players. Sampling frequencies such as 22 KHz result in frequency responses closer to AM radio. An 11-KHz sampling rate results in a frequency response slightly less than that of a typical telephone.

The fidelity with which the digital audio data samples can follow the original analog waveform is determined primarily by the number of data bits in the sample. An 8-bit audio sample digitizes 128 positive and 128 negative steps for a total of 256 different levels. Sampling using 16 bits increases the number of levels to more than 64,000. The number of sampling levels determines the dynamic range of the sampler. Sampling with 8 bits is acceptable for speech with carefully set recording levels, but music tends to require more dynamic range with lower noise and is best accomplished with 16-bit sampling. The trade-off of sampling rate and sampling size for digital audio is driven by bandwidth (number of sampled data bytes per second) versus fidelity (music or voice). Stereo audio sampling requires twice as many bytes as mono audio.

Digital audio signal-processing techniques can be applied to the raw digital audio data to compress the size. Adaptive Delta Pulse Code Modulation (ADPCM) is one of the digital compression techniques which can yield a 4X reduction in digital audio data and is based on the principle that digital audio data are not random, but instead each sample value tends to be close to the last sample. ADPCM compression allows high-fidelity stereo digital audio to be delivered in the same bandwidth as lower-frequency stereo 8-bit data, as shown in Figure 6.2.

The IMA has a standard ADPCM algorithm for digital audio compression which is available at no charge for any multimedia platform, for audio data on any medium such as CD-ROM, floppy, or network. Sony has developed an ADPCM algorithm called CD-ROM/XA that is available only on CD-ROM/XA players from Sony and is licensed for use on the Philips CD-I player and some CD-ROM drives. Even with digital audio compression, the size of audio data will preclude the use of low cost storage-like floppy disks from widespread interactive multimedia use. Some of the emerging multimedia game software requires more than 10 high-density (1.2 or 1.44 MByte) floppies to be copied to the personal computer hard drive before playing the game.

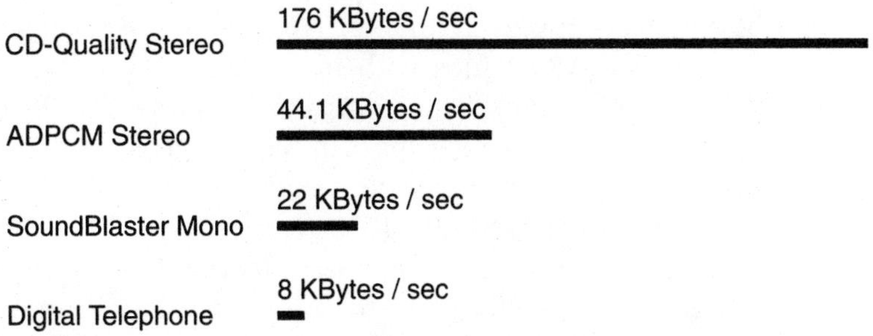

CD-Quality Stereo
176 KBytes / sec

ADPCM Stereo
44.1 KBytes / sec

SoundBlaster Mono
22 KBytes / sec

Digital Telephone
8 KBytes / sec

Figure 6.2   Multimedia digital audio bandwidth requirements

Another form of digital audio compression has been developed by the music industry called MIDI (Musical Instrument Digital Interface). MIDI data give just the frequency and duration of notes and which instrument (track) is involved. MIDI data result in whole songs being compressed to tens of KBytes. One drawback is that the mapping of instruments is not standard, so MIDI playback on various boards may sound quite different. Even with the General MIDI standard specifying common instrument mappings for MPC machines, the actual quality of reproduction of a standard instrument like a piano varies widely and still results in dramatically different sounds with different synthesizers.

MIDI also specifies a serial interface for control of devices like electronic organs and music synthesizers. Inside some early electronic organs were music circuits using a technique called FM synthesis. An FM synthesizer is basically a digitally controlled oscillator circuit. One of these circuits was used on the AdLib sound board for the PC and found quick acceptance by the game software manufacturers. MIDI data can be played on an FM synthesizer with the proper software device driver.

Another technique for digital music is wave-table synthesis. This technique has been used on high-end electronic music synthesizers and was pioneered on PC Compatibles by some clever shareware programmers on the Amiga. The Sound-Tracker program used the Amiga four-channel DMA (Direct Memory Access) sound circuitry provided four-track 8-bit sampled music synthesis. Only one note for each instrument is sampled, and the computer calculates different notes and volume control per the script, which also stores the music samples in a table. This technique results in music files called MOD files with from tens to hundreds of kilobytes, but the music is identical when played back on different machines because the sampled data is part of the file. Also digitized voice can be incorporated in the file (not yet possible with standard MIDI synthesizers). This method of music synthesis has been ported to the PC (Trakblaster and others) and directly plays the MOD files from the Amiga on various sound boards available for PC compatibles.

Personal computers are now capable of decompressing digital audio ADPCM data and running sampled music synthesis programs using the main microprocessor as a background task in addition to running other multimedia tasks such as displaying graphics or video animations.

Digital Graphics Data and Compression

Graphics data comprise both digitized and computer-generated still images. Image scanners capture either gray scale (typically one byte for 256 shades) or true color (typically three bytes representing 256 shades each of red-green-blue, RGB data). Since the PC VGA and the Mac Video Adapter use color lookup table (CLUT) circuitry to map 16 or 256 table entries to 16 million (true-color) space, the scanned images need to be converted to 8-bit CLUT format. This conversion results in a fixed compression ratio of 3X over the original 24-bit per pixel image, as shown in Figure 6.3.

The memory to store a single full-screen graphics image ranges from 921 KBytes for a true-color 640 × 480 image to 307 KBytes for an 8-bit CLUT image. Several other compression techniques have been used to reduce the storage requirements for graphics. Lempel-Zev-Welch (LZW) lossless compression of CLUT graphics such as the CompuServe GIF format can reduce the image size to about one half of the

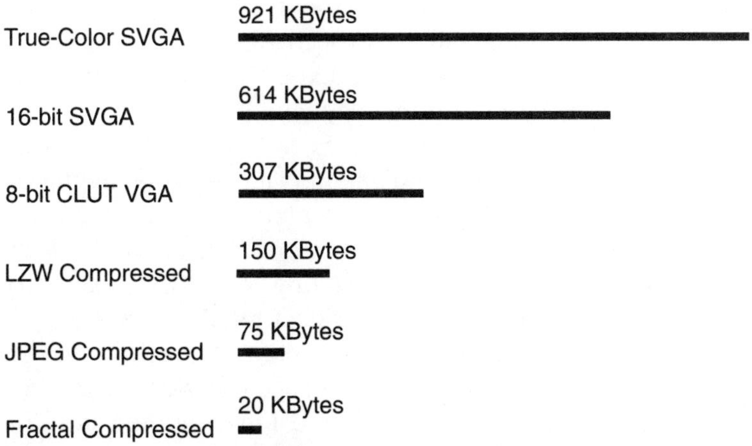

True-Color SVGA
921 KBytes
▬▬▬▬▬▬▬▬▬▬▬▬▬▬▬▬▬▬▬▬▬▬▬▬

16-bit SVGA
614 KBytes
▬▬▬▬▬▬▬▬▬▬▬▬▬▬▬▬

8-bit CLUT VGA
307 KBytes
▬▬▬▬▬▬▬▬

LZW Compressed
150 KBytes
▬▬▬▬

JPEG Compressed
75 KBytes
▬▬

Fractal Compressed
20 KBytes
▬

Figure 6.3   Multimedia graphics storage requirements

CLUT image. Lossy compression schemes such as JPEG (Joint Photographic Experts Group) can compress true-color images about 12X (approximately one-half the size of the same image in GIF compressed format). Another image compression technique is fractal compression. This technique can result in fifty times the compression of the original true-color image. Images painted using a computer tend to be more limited in color variations and may be compressed without loss more than ten times by techniques like GIF or run-length encoding (RLE). Lossy compression can be accepted in many cases for lifelike pictures, as contrasted with graphics or text where the artifacts of lossy compression may be unacceptable.

Digital Video Data and Compression

Video data can be considered a stream of graphics images with the data rate corresponding to the number of bytes per image multiplied by the number of images (or frames in TV terminology) per second. For example: 640 × 480 true-color uncompressed images displayed at 30 frames per second require a sustained data transfer rate of over 27 MBytes/sec. Using JPEG compression for those images would reduce the transfer rate to around 2 MBytes/sec, as shown in Figure 6.4. As a reference, the CD-ROM standard data rate is 150 KBytes/sec and computer networks like Ethernet and Token-Ring are capable of only about 1 MByte/sec or 2 MBytes/sec theoretical maximum, respectively.

MPEG (Motion Picture Experts Group) compression takes advantage of the principle that most moving pictures contain a significant amount of area that does not change from one frame to the next and attempts to encode only the difference between frames (objects in motion) to achieve compression ratios approaching one hundred times over the true-color source material. The resulting 270 KBytes/sec would still exceed standard CD-ROM data rates and would represent a significant fraction of a standard network bandwidth. Two short-term approaches to reduce the video data rate requirement are smaller images and lower frame rates. To reduce the video image size from 640 × 480 to 320 × 480 is a two

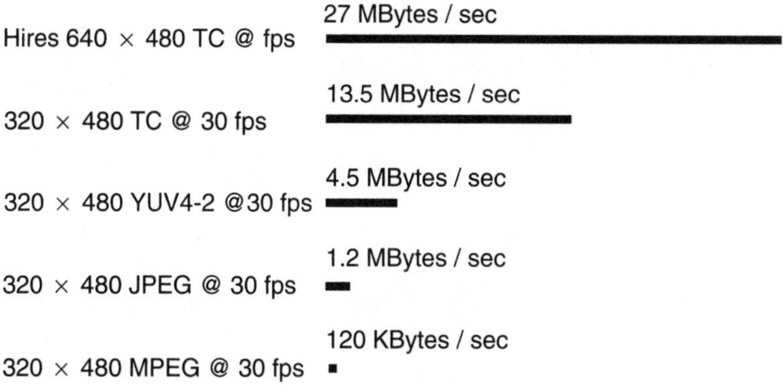

Hires 640 × 480 TC @ fps

27 MBytes / sec

320 × 480 TC @ 30 fps

13.5 MBytes / sec

320 × 480 YUV4-2 @30 fps

4.5 MBytes / sec

320 × 480 JPEG @ 30 fps

1.2 MBytes / sec

320 × 480 MPEG @ 30 fps

120 KBytes / sec

Figure 6.4   Multimedia video bandwidth requirements

times reduction. Another two times of additional reduction can be achieved at 15 frames per second compared to 30 frames per second.

The Role of Delivery Media (CD-ROM and Networks)

Multimedia content will be delivered to users in a variety of ways (media) including CD-ROM, networks (both wired and wireless) and other techniques under development. Today, CD-ROM represents the most cost-effective delivery medium with over 600 MBytes of digital data on a CD for well under two dollars in production costs. This medium is well suited for one-to-many multimedia content such as entertainment, education, and training material that remains static for at least a couple of months. Data that change sooner than that are probably better suited for some kind of network delivery mechanism, including wired LANs (local area networks such as Ethernet and Token-Ring), WANs (wide-area networks using leased or dial-on-demand telephone connections at data rates of 2400 bps through T1 1.5 Mbps and ISDN) and emerging wireless networks such as digital cellular, radio/TV broadcast, and others.

Multimedia data such as videoconferencing require two-way connections of continuous data streams that require networking. Movie multimedia content, on the other hand, requires storage on a cost-effective medium like CD-ROM or other optical media such as MO (magneto-optical read/

write) or WORM (write-once read-many), but still might be either directly delivered to the customer by VideoCD purchase or rental or delivered over any of the networks mentioned above. Design considerations for network delivery become much more complex due to multiuser bandwidth partitioning issues. If the bandwidth cannot be guaranteed, then very complex algorithms are required to minimize the impact of data dropout on both the multimedia sound and graphics or video information streams. Estimating the computer storage requirements of multimedia data is a straight-forward calculation multiplying the instantaneous multimedia data rate by the number of seconds of the data length for each stream of data and then adding the size of all the streams, as graphically illustrated in Figure 6.5.

Multiple streams of multimedia data such as a movie including one audio and one video stream can practically be delivered only as one interleaved stream of digital data. Interleaving is an approach that mixes two streams of data into one stream. The data are mixed into the interleaved stream such that each of the independent stream data rates are maintained.

As an example, if a movie is to be delivered on a standard CD-ROM, then 150 KBytes per second is the maximum total continuous data rate available for the interleaved data. Due to the slow seek time of CD-ROM mechanisms, that data rate is also the minimum to maintain streaming. The digital audio and video data then must total 150 KBytes per second. If the audio data are 16-bit sampled stereo, compressed using ADPCM at a sample rate of 22 KHz, they require 22 KBytes

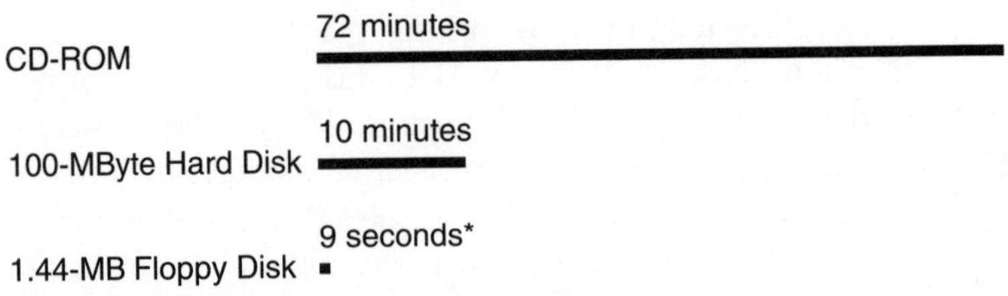

CD-ROM   72 minutes

100-MByte Hard Disk   10 minutes

1.44-MB Floppy Disk   9 seconds*

*The floppy drive is capable of less than 30 KByte/sec streaming data rate

Figure 6.5   Multimedia delivery media capacities (MPEG-I Data @ 150 KBytes/sec)

per second. This leaves 128 KBytes per second for the video data. Using techniques such as MPEG compression for the video, which do not result in the same number of compressed bytes per frame, the interleaved audio+video data must be padded with additional null data which is discarded on playback. Sometimes, multiple audio tracks (foreign languages) might be desired, interleaved with a single video stream. The design implication of this is that less video bandwidth is available.

The example above of a multimedia movie with interleaved data can be calculated for network data delivery by replacing the 150 KBytes per second guaranteed data delivery rate of a CD-ROM with the data delivery rate available for the network under consideration. Current computer networks do not guarantee data delivery rate, and so efforts to compensate for data loss are under investigation.

Scripting Languages

Up to this point we have mentioned compression of video, audio, and graphics data as a method to minimize the impact of the enormous size of multimedia data upon content delivery and storage systems. Playback and creation of multimedia data have different hardware requirements, which could necessitate different platforms (with the exception of videoconferencing, where the data are created and another stream is played back simultaneously). We first consider the computer architecture trade-offs for multimedia playback.

Playback of multimedia data includes not only simply playing audio, graphics, and video data but also handling navigation and interaction with the content in accordance with an authored script. The execution of the script and operation of the user interface (keyboard, screen, mouse, audio, touchpad, and others) is typically handled by the microprocessor in the personal computer, in conjunction with hardware input/output interfaces. These interfaces may actually include processors (keyboard controllers, mouse controllers, CD-ROM controllers) in their own right, but in some subsystems such as graphics and audio, there may not be any processor, in which case the microprocessor interfaces with the subsystem directly at the raw data level (audio samples or graphics bytes).

A script language provides an abstraction of how the user input devices and the multimedia audio and video subsystems operate, allowing multimedia platforms to be designed with different amounts of hardware assistance and relative independence from the actual hardware components selected. There is not a standard script language available for interactive multimedia, so current multimedia applications are forced to use custom languages or those provided by different multimedia development toolkits. Lack of such a standard script language is an authoring barrier for multimedia titles.

Publishing Formats

Multimedia content is being published in a variety of formats. Floppy disks are still the most widely available distribution mechanism for personal computers and some of the multimedia content is being delivered in this form, such as the Knowledge Adventure series for PC compatibles. Because of the limited storage capacity of floppy disks, these packages tend to be titles like multimedia audio clips and entertainment packages with limited audio and no video. Floppy disk multimedia content is predominantly either PC-compatible MS-DOS format or Macintosh format.

CD-ROMs are the other dominant publishing format for multimedia content because of their very large data capacity and inexpensive publishing costs leveraging CD-audio production equipment. The data format on CD-ROM is predominantly ISO-9660, which is an international standard specifying minimum directory structures and does not specifically address multimedia data.

Two new CD-ROM formats directly address multimedia data. CD-ROM/XA format defines a specific way to combine ISO-9660 digital data with interleaved multimedia data. The digital audio data are compressed four times using the CD-ROM/XA ADPCM algorithm and interleaved with either more audio tracks or other multimedia information. CD-ROM/XA playback requires compatible CD-ROM drives and compatible digital audio hardware currently limited to a couple of Sony handheld players, CD-I machines, and PC compatibles with a special Sony CD-ROM drive and Sony digital audio option board. Kodak has defined another multimedia CD-ROM format called Photo-CD. This format defines special Photo-CD graphic images and will include digital audio in a future release. Photo-CD also includes the ability to add pictures and files to a recordable CD, but this feature is not available in all CD-ROM players. There is currently not a multimedia publishing format which can be played across different kinds of personal computers. This cross-platform multimedia format is one of the goals of the IMA.

Applicable Standards

De Facto, Consortia, and Formal

Standards reduce the costs for manufacturers and consumers. The presence of competition both keeps prices for platforms reduced and increases the variety of content available. Standards include de facto (PC compatibles, ISA bus, GIF files, MOD files), consortia (EISA, MCA, VL-Bus, PCI, IMA, MPC, General MIDI) and formal (JPEG, MPEG, ISO-9660). The work of trade associations like the IMA toward developing cross-platform multimedia standards is based on the assumption that increasing the number of multimedia platforms will greatly encourage the development of multimedia titles.

Bus Standards

Personal computer bus standards such as ISA, EISA, MCA, NuBus, VL-Bus, and PCI provide standardized interfaces for the design of multimedia authoring and playback hardware boards. Each of these busses provides the ability to attach digital audio and digital video hardware to personal computers. These busses differ in terms of how fast digital data can be transferred, the physical size of the board, and how many boards can be plugged into one personal computer. For the user, the availability of particular multimedia devices and their price is almost directly related to the installed base of personal computers with a particular bus architecture.

The ISA bus is available in over 70% of the personal computers available today and has the widest variety of multimedia options available. The other busses have some technical advantages over the ISA bus, but current multimedia platforms like the MPC do not require more than ISA bus capabilities.

Multimedia Data Standards

Audio compression standards such as the IMA ADPCM, MIDI, and music synthesis (FM or MOD wavetable) are important requirements to encourage the widespread development of multimedia content. Graphics compression standards such as RLE, GIF, fractals, JPEG, and video compression standards like MPEG are critical to the success of multimedia.

Delivery Standards

Current multimedia content delivery is centered on CD-ROM with ISO (9660) format. For PC compatibles, Microsoft RIFF format, which encapsulates text, audio, and video for playback under Windows 3.1, is the most popular format. For the Mac line of computers, multimedia content is typically Quick-Time. Photo-CD and CD-ROM/XA formats are beginning to emerge.

Platform Standards

Several unique platforms for playing multimedia data have appeared, including Philips CD-I and Tandy VIS, both of which attach to a color TV and have handheld game-style controllers. The MPC is a consortia-defined open platform centered on a 386 PC with digital audio, CD-ROM, and joystick.

Infrastructure

Computer Architecture

Basically any type of microprocessor can be used in a multi-media platform, including CISC (complex instruction set) Intel X86 or Motorola 68XXX, RISC (reduced instruction set) Alpha, MIPS, or PowerPC. The computer architects select various elements of a personal computer based on anticipated use. For a limited purpose machine such as the CD-I and VIS designs, the architects balance the cost of every subsystem based upon their intended use. Extra capability (processor performance, memory capacity, upgradability, and option support) is typically minimized to meet cost goals.

Architects of general-purpose personal computers usually select elements to maximize the number of uses for the machine, including some capabilities (processor upgrade sockets, memory sockets, option board sockets) that add cost to the basic system but provide broad capabilities to expand the platform with add-in options. With the current evolution of multimedia formats and standards, more flexible platforms are a good investment.

Code Libraries/Upward Compatibility

Software compatibility is an issue for both content developers and multimedia users. Multimedia tools and libraries predominate on personal computer platforms like the PC compatibles and the Mac. This results in more multimedia content availability for those platforms. Significant expansion of the current multimedia market is caught in the chicken-and-egg quandary. Due to the significant costs for developing multimedia content, developers await a large number of platforms capable of playing titles they wish to create.

Personal computer manufacturers have been unwilling to add significant hardware features and corresponding costs (digital audio and digital video) in a price competitive environment without significant market demand and the availability of large numbers of multimedia titles. Therefore, today's consumer interested in multimedia may choose a limited-flexibility platform with an attractive price tag (CD-I, VIS, etc.), a very flexible platform with a realtively high price tag (MPC, etc.), or upgrade an existing personal computer with multimedia attachments.

Regulation and Legal Constraints

Limitations on Collaborative Activities

The Federal Trade Commission (FTC) enforces U.S. government trade regulations. Industrial collaboration not occurring in open public forums, such as trade associations, is scrutinized for potential restraint of trade violations. The MPC specification was developed in a small consortium and requires licensing for conformance. The MPC specification is applicable only to PC-compatible architecture. The Cross-Platform Multimedia Project of the IMA is being conducted in an open trade association forum including a broad base of companies representing both hardware and software developers and some end-users. The IMA goal is to develop a combined installed base of multimedia platforms to stimulate more multimedia development than any one platform is able to engender alone.

Outlook

Higher Horsepower Means More Multimedia

Due to the very large physical size of multimedia data streams it is typically impractical to consider off-line decompression of the data into raw form for playback. Therefore, the following discussion assumes real-time operation maintaining continuous streams of incoming compressed data (that may be interleaved) and continuous output of uncompressed data (to sound systems or video display systems, for instance).

Decompression of multimedia data on a particular computer playback platform can be accomplished with a number of architectures that can be divided into either hardware or software approaches. For instance, decompression of audio data in real time using ADPCM algorithms is being accomplished daily without difficulty by personal computers using software. However, graphics decompression routines using the main CPU have taken several seconds per picture until recently. A variety of hardware solutions have appeared on the market which dramatically speed up the decompression of graphics images. Some hardware solutions to decompressing video have also appeared.

Multimedia platforms can be constructed with personal computers with moderate performance assisted by specialized hardware to handle decompression and playback of digital audio and video, or they can be constructed from more powerful personal computers performing the audio, graphics, and video decompression with low-cost hardware using software decompression techniques. With the dramatic increases in the power of personal computers without increasing prices, the software-based decompression multimedia platforms offer more flexibility as newer compression techniques emerge for both audio and video compression.

Multitasking

Multitasking is the ability of a computer to operate on a number of tasks concurrently. With a single processor in a multimedia playback platform, DMA (Direct Memory Access) hardware moving audio and/or video data is required to maintain the continuous streams of multimedia data. Multimedia platforms with audio or video processing subsystems off-load the main personal computer but may require more complex control to synchronize the audio and video multimedia data.

Most computers simulate the ability to perform multiple tasks simultaneously by rapidly switching from one task to another, assuming that any one task does not require the continuous capability of the main processor. When this multitasking operation maintains carefully timed operation, it is known as real-time multitasking. The need for real-time multitasking is evident in observing some of the video approaches taken in current multimedia titles. Synchronizing the interleaved digital audio to the digital video data streams on playback is a requirement for lifelike movie sequences.

More Consistent Set of Functionality and Media

The potpourri of multimedia choices confronting the multimedia content authors is a mixed blessing. The cost of authoring multimedia data is high due in large portion to its size. Selection of a format for the data today typically limits the available market for that data to one platform. Agreement on a more consistent set of standard functionality for multimedia platforms and the media for the content delivery should eliminate a couple of the barriers to more widespread multimedia availability.

7

Computer Software

JAMES GREEN

Software to run interactive multimedia applications on computers is obviously linked tightly to the computer's architecture and hardware design. Another view is that computer software, which is usually developed and sold after the hardware product is shipped and installed, creates the opportunity to "redesign" the product after it has been sold. This concept, which is unheard of in consumer electronics circles, creates entirely new "aftermarket" sales (that is, products sold to add onto an existing installed base of products after they are initially sold). Software's ability to reengineer products after they have been designed is directly responsible for today's home multimedia computer market.

As with other industries, there are different optimization points for business and consumer applications which take into account user expectations (to the extent that home computers can actually fully meet consumer entertainment requirements). General business users, specialized business users, groupware users, and media consumers are contrasted in this chapter, illustrating divergent requirements among the various groups.

Multimedia and computers is not new, but all-digital multimedia data types have only recently been feasible due to bandwidth, speed, and storage media limitations. These constraints are falling away and are exposing a new set of issues and limitations that this chapter discusses. Analog multimedia, the forerunner of current systems, required only that the computer software exert control over media streams. Digital audio and video, however, require the computer to manage both control and media streams—a much more computer-intensive job.

Multimedia frequently means the processing of multiple, parallel data streams. Current popular desktop-computing operating systems were not designed to manage this kind of task. This chapter reviews the issues associated with real-time synchronization of multimedia data types and the resource management implications. For example, seemingly simple problem of achieving "lip sync" between audio and video (i.e., the proper synchronization of sound and video) turns out to be a major system-design challenge that cannot be taken for granted by systems or applications designers.

Multimedia computers must be "taught" by software to play multimedia, and they are sometimes unwilling students because of their hardware history. Thus the era of the "architected" environment was born. After years of less successful attempts, designers of system software struck upon a more systematic way of designing software to manage multimedia tasks. This chapter reviews the evolution of such designs and provides a reference framework for various multimedia structures. The need to separate control from data flow, among other things, is highlighted.

Multimedia software, like each topic in this book, is a broader topic than can properly be covered in one chapter. For one thing, software exists in at least three distinct categories as it relates to multimedia: consumer titles, developer tools, and system software. These are not simply separate categories; they are totally separate markets. It is certain, how-

ever, that the future of interactive multimedia is more than dependent on software—it is software.

History

The Beginnings of Multimedia

Beginnings are difficult things to pin down. Histories always have a point of view and selecting one can be difficult. No matter which event is chosen as the defining moment, there are other points of view that would suggest still earlier ones. This is particularly true for multimedia since it represents, on many different levels, a synthesis, or to use the currently popular phrase, a convergence. In other words, several threads of development from previously unrelated areas are coming together in some new way. A history which strives for completeness would presumably have to cover all these threads. This would be a formidable task and completely beyond the scope of the present work. However, a brief review of some of the choices may serve to illustrate the point.

In 1945, Vannevar Bush published an article entitled, "As We May Think" in *Atlantic Monthly*. In this article, he describes the following scenario:

> One can now picture a future investigator in his laboratory. His hands are free, and he is not anchored. As he moves about and observes, he photographs and comments. Time is automatically recorded to tie the two records together. If he goes into the field, he may be connected by radio to his recorder. As he ponders over his notes in the evening, he again talks his comments into the record.[1]

One could argue that the system he is (rather prophetically) describing is a multimedia system, and then proceed

---

1. Vannevar Bush, "As We May Think," Reprinted in *Computer-Supported Cooperative Work: A Book of Readings*, Edited by Irene Greif, Morgan Kaufmann, 1988, p. 24.

to present nearly the entire history of computing. This could follow the sequence of events and inventions that gave rise to database management, computer networking, electronic mail, computer conferencing, and on-line random access storage, and from there, branch off to explore still emerging *groupware* applications such as videoconferencing, shared workspaces, and various other forms of *computer-supported cooperation*.[2]

Another thread could begin with the work of B. F. Skinner and others in the 1950s on "programmed instruction" techniques for use in education and training. This could show how this work ultimately led to development of Computer-Aided Instruction (CAI) techniques. It could then go on to describe how an early concern of CAI developers (i.e., how to take advantage of existing linear videotape training materials) led to the introduction of Interactive Video Systems which used interactive videotape and videodisc players interfaced to microcomputers.[3] This would ultimately lead to a discussion of the work on Digital Video Interactive (DVI), conducted first at the David Sarnoff Research Center, and later at Intel Corporation,[4] and from there to the recent explosion in digital audio and video products for desktop computers.

One could reach even further back by selecting a particular form of media such as *Print*, and then follow its evolution as publishers embrace new technologies, from the printing press through desktop publishing systems. The next items along this thread might include such topics as electronic publishing, digital imaging, and the CD-ROM. Other threads that might prove interesting include an examination of the effects of digital computer systems on film and video production, which are replacing traditional technologies in

2. Irene Greif (Editor), *Computer-Supported Cooperative Work: A Book of Readings*, Morgan Kaufmann, 1988.
3. Steve and Beth Floyd, *Handbook of Interactive Video*, Knowledge Industry Publications, 1982.
4. Arch C. Luther, *Digital Video in the PC Environment* (Second Edition), McGraw-Hill, 1991.

such applications as video editing and animation. Yet another thread might show how interactive video games are affecting our views about entertainment, and then proceed to make the case for virtual reality[5] and cyberspace.[6]

By now it should be clear that multimedia has diverse roots. It is primarily due to the number of distinct points of view, that the term "multimedia" has been so difficult to define. The perspective of this chapter is that of computer software, therefore beginnings occur when programmable digital systems are first used to solve a problem or automate a process. The "media" part of it came in when digital systems began to be used for the manipulation of information in the form of audio, video, and photorealitsic images. That has occurred in several places and times in the recent past, and in a wide variety of industries, each with different goals.

Nevertheless, the use of programmable digital systems follows a typical adoption path. First, specific parts of existing processes are automated. (This can sometimes evolve into a fully automated process.) The next phase is characterized by a redesign of the process itself so that it is more appropriately tuned to the digital technologies employed. And the third is the emergence of new processes that exploit the characteristics of programmable digital technologies, to provide additional benefits. Depending on the speed of adoption and the degree of penetration of digital systems from one industry to another, multimedia systems may exist in any of these three phases. As a result, our discussion of the state of multimedia software will find us jumping around from one thread to another (often picking them up in the middle) as multimedia software applications emerge and evolve in different problem domains. We will also examine the effects digital technologies have on these problem domains, once they have been introduced. And finally, we will discuss some of the ramifications of the convergence of multiple problem domains into a single solution space, namely that of programmable digital systems.

5. Howard Rheingold, *Virtual Reality*,
6. Michael Benedikt (Editor), *Cyberspace: First Steps*, MIT Press, 1991.

Interactive Videodisc Systems and Applications

Due to the importance of video and audio to most definitions of multimedia, and because we must start somewhere, we will begin by looking at the first attempts to provide access to, and control of, audio-visual materials using a videodisc player connected to an microcomputer system. Laser videodisc technology has been around since the early 1970s. There are two primary formats, CLV (for constant linear velocity) and CAV (for constant angular velocity). CLV is used primarily as a high-quality distribution medium for home video enthusiasts. CAV formatted discs provide a number of features that are not available on CLV discs, such as picture stops, freeze frame, step frame, slow motion, and frame searches, making them more suited to interactive applications than CLV. Laserdisc players are, themselves, classified by levels of interactivity beginning with Level 0 (linear playback only) to Level III (which include interface ports which enable them to be controlled by an external computer).[7] CAV videodiscs and Level III videodisc players interfaced with interactive microcomputers enabled developers to create some of the first interactive multimedia applications.

These applications use the computer to direct the activities of the user, control the videodisc player, and record any results. For example, in a training application the computer might log the student's name, begin the session at the appropriate chapter on the videodisc, stop the videodisc at specific intervals and ask questions, record the answers, calculate a grade, and decide whether to proceed to the next chapter or review previous material. The computer could also perform other tasks such as keeping track of which questions were missed most often. This would enable the application developer to revise unclear material or add additional sections to further explain or illustrate the point. Such applications are

---

7. Steve Lambert and Jane Sallis (Editors), *CD-I and Interactive Videodisc Technology*, Howard W. Sams & Co., 1987.

typically referred to as *Computer Based Training* or CBT applications.

Another common use of interactive videodisc technology is the Information Kiosk, which can be seen in various high-traffic locations such as airports, museums, and shopping malls. These applications can provide tourists with hotel, restaurant, or sightseeing information, locate items of interest for museum visitors, or provide electronic catalogs for shoppers. These kiosks are generally placed in enclosures resembling automated-teller machines, and incorporate touch screen displays and user interfaces that let the user browse the data via a series of simple menu selections. Interactive videodiscs are still widely used in these kinds of applications because the infrastructure for production and application development is well established.

Although computer-controlled videodiscs are among the first successful examples of interactive multimedia, their use differs from most computer processes. Computers control videodisc players by issuing commands such as *play, stop, seek frame number,* etc., but do not process the video itself. Video segments or images from the computer are combined with the video by overlaying them on top of the video pictures from the disc. This combining process occurs on what is called a graphics overlay card, which does not alter the video from the disc itself. Videodisc signals are stored on the disc using the analog NTSC television standard—the same standard used for television broadcast and VCRs. So, the computer does not actually process the video media stream at all, thus permitting fairly low-powered computers to exert a high amount of control over the video presentation without creating a computational burden to the computer or its data paths. With the advent of more powerful computers and larger digital storage media such as CD-ROM, the shift from analog to digital is now under way. Nonetheless, analog laser discs still provide the highest quality video for interactive multimeida applications available today, although that is likely to change when new video standards such as MPEG II and high-density compact discs become available.

Binary Representation of Audio and Video

Sounds and images in nature are analog. They are made up of continuous fluctuations in air pressure, or continuous flows of electromagnetic radiation over a (fairly narrow) frequency range known as visible light. Videodisc and videotape systems store audio and video in analog form. Until recently, these systems were all based on technologies that converted analog sound and light into fluctuating electrical signals for storage and transmission, and then back into sound and light for the benefit of some human listener or viewer. It is also possible to do this with digital technologies, where the analog sound and light is converted into a string of binary digits, rather than electrical impulses. This process is called an *analog-to-digital conversion*, and it has two fundamental steps: *sampling* and *quantizing*. Sampling involves the measurement of the analog signal at specific regularly spaced points in time. Quantizing involves the assignment of an appropriate value to represent the value of the signal at that point.

Figure 7.1 illustrates the simple technique of taking the value of the signal at each clock pulse and holding it until the next clock pulse. It is not difficult to see that the resulting digitized data do not represent the analog signal exactly, but are rather an approximation. The goal of a good analog-to-digital conversion system is to approximate the original signal as closely as possible. This can be done by sampling more often, by using more bits to store more accurate quantization values, or by using a more sophisticated method of assigning quantization values (by using a weighted average calculation, for example). The limits of human perception of light and sound place some bounds on this problem by allowing us to ignore information that we cannot perceive anyway. The end result of this process is a string of binary digits which represent the original audio or video signal.

Once in digital form, audio and video become *data types*. A data type is simply a digital representation of some particular kind of data, such as letters or numbers. Let's describe the concept of a data type by way of illustration. Consider the letter "J", which is nothing more than a code that repre-

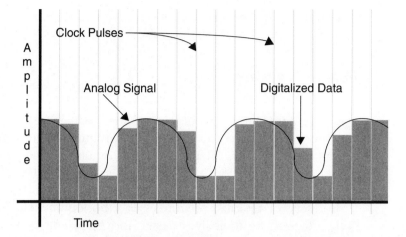

Figure 7.1 The value of the amplitude of the analog signal is taken as the time of each clock pulse (sampling). This value is then turned into a binary number (quantizing). Notice that whereas the analog signal is changing continuously, the digital signal maintains its value for the duration of time between clock pulses. Good analog to digital conversion systems use various techniques in an attempt to minimize the differences between the analog and digital versions to the limits of human perception.

sents a sound (which, if this were a multimedia book, I could render for you). When encoding this letter using binary digits, we could simply choose a number, and agree that this means "J" (which is in fact what the ASCII committee actually did). Now we can build computers and write software that interpret the action of pressing the letter J on a keyboard with the storage of the binary sequence (the encoding) which, when sent to the computer's display, produces a reasonable representation of the printed letter J (the decoding). In common computer terms, letters like J are data, and data are of different types.

The term *media* is often used to refer to a means of mass communication, as in the print media, or the medium of television. Since all these types of media carry information in different forms (i.e., text or video) the idea of encoding these different media types as computer data has given rise to the term *multimedia*, even though they are perhaps more accurately referred to simply as data types (which is they way I will refer to them throughout the rest of this chapter).

The Role of Software

It has been a dream of mankind since at least the beginning of the industrial age to create a universal machine, capable of performing any function the user requires, automatically and expeditiously. Such a machine would be the ultimate tool, the culmination of tens of thousands of years of tool-making. It would presumably free its creators from the drudgery of tedious, repetitive, or laborious activities, and therefore make their lives easier and more enjoyable. Well, we have yet to create such a marvelous machine. We have made a great deal of progress toward this goal, however, particularly in the last fifty years or so since the creation of the first general purpose computers.

The central idea that made such general purpose machines possible is *programmability*; the ability of the machine to have its behavior altered based on a set of instructions provided by the user. Today these instructions are called computer *software*, but the notion of programmability actually predates the computer. In 1805, Joseph Marie Jacquard introduced a system for weaving cloth which enabled the same machine to produce any number of patterns. He accomplished this by building a loom that could be reprogrammed via punched cards. Charles Babbage, one of the legendary forefathers of the computer era, picked up on this technique when he designed his Analytical Engine, which incorporated two sets of cards, one to carry the data and the other to carry the instructions, or operations to be performed on the data. Although he never finished building it, he quite clearly intended the Analytical Engine to be a general purpose computing device.[8]

In the years before 1945, a number of pioneering efforts were undertaken in pursuit of Babbage's dream, and a full

---

8. Herman H. Goldstine, *The Computer: from Pascal to von Neumann*, Princeton University Press, 1972, pp. 19–21.

survey of these efforts would be (as has already been mentioned) beyond the scope of the present work. However, a major refinement was introduced in 1945 that has a direct relationship to our topic. It was in that year that John von Neumann published the *First Draft of a Report on the EDVAC*, which included "the essentials of the first programming language for a stored program computer."[9] EDVAC, which stood for Electronic Discrete Variable Automatic Computer, had a much larger memory than had previously been feasible, enabling it to store both the instructions and the data in the computer's memory rather than on a read-only medium, like punched cards. This required the instructions to be encoded in a machine readable form, which in binary computer systems is a sequence of zeros and ones corresponding to the on/off states of the computer's circuitry. Rather than force human programmers to memorize these binary sequences, a higher level language was developed which was easier for humans to read and write, but could easily (and automatically) be translated into the machine's binary code.

This automatic translation step involved the use of the computer as an aid in developing computer software for the first time. Since then, as computers have grown less expensive and more powerful, a number of software products have appeared that are designed to support the software development process, including editors for writing programs (also called source code), compilers for translating the source code into machine code, debuggers for locating and identifying errors, and version control software for keeping track of the source code as it changes and to control access when more than one programmer is working on the same program. Today these tools are commonly used by developers of all kinds of software, including multimedia tools, applications, and titles. Continued development of these tools is important if computers are to find their way into the lives of nonspecialists, since people tend to think like peo-

---

9. *Papers of John von Neumann on Computing and Computer Theory*, Edited by William Aspray and Arthur Burks, MIT Press, 1987, pp 145–149.

ple, not like binary computers. People generally want to use natural (i.e., human) languages, not computer languages. And people would like to be able to communicate with their computers the way they communicate with themselves, using speech and gesture, rather than typing commands at a keyboard. Increasingly sophisticated hardware and software, in particular hardware and software that support multimedia will inevitably make this possible. In the introduction to their book, *Multimedia Interface Design*, Roger Dannenberg and Meera Blattner concluded as much when they said,

> As we continue to develop computers along lines that reflect our needs, we find it important to interact with computers using all of our senses and communication abilities. Why should we settle for less? The use of audio, video, graphics, and animation in computer systems is not simply an extension of more conventional computers; rather, it is an attempt to complete the "universal machine," to develop computers that can more fully understand and communicate with human beings.[10]

Summary

So from a computer software perspective, the key technical elements of a multimedia system are programmability and the incorporation of audio and video as digital data types. And like all digital systems, multimedia software must solve a problem or facilitate a process for the user of the system, and it must do so in a way that is more economical than the previously unautomated system, or it must provide some additional features desired by the user. It is important to understand that different users have different needs and come from different problem domains. Computer software is often designed and developed to fit into an existing (and often specialized) work environment which has its own methods, terminology, and culture. The inevitable result is that behavior of software products can vary considerably from one to another, even though they may be performing

---

10. Meera M. Blattner and Roger B. Dannenberg, *Multimedia Interface Design*, ACM Press/Addison-Wesley, New York, 1992, p. xvii. (From the introduction.)

similar tasks. This is an important point to keep in mind as we survey the various users of multimedia systems in the next section. It is also important to keep the idea of a "universal machine" in there as well.

Profile of the User

All good software systems are created with a target user in mind. By profiling the intended user, a software developer can make assumptions about the knowledge and experience the user will bring to the task. This analysis begins by asking such questions as: What problem is the user trying to solve? What are the key features the user wants the system to have? What kind of user interface will the intended user find intuitive? What level of expertise will the intended user have in performing the task for which the software system is to be used? Is the user a specialist with specialized knowledge of the task, or a casual user with only a rudimentary amount of knowledge? What is the user willing to pay for the system? For multimedia systems, additional questions might include: What level of quality is required as an outcome? Will the final product be used as an internal presentation to management? To customers? Or will it be used in a national advertising campaign or perhaps the next Steven Spielberg movie?

There are many ways to categorize users of multimedia systems. For the sake of simplicity, we will examine the following general categories: General Business User, Specialized Business User, Groupware User, Media Producer, and Media Consumer. For each category we will characterize the

Table 7.1   User Categories

General Business User	Those who use desktop computers
Specialized Business User	Computer-aided multimedia creation
Groupware User	Collaborative work at a distance
Media Producer	Computer-generated effects and editing
Media Consumer	Mass market consumer (games/edutainment)

user and describe some of the ways in which multimedia software is being used.

General Business User

The category *general business user* includes nearly everyone who uses a desktop computer as part of his or her job. Users of multimedia systems in this category tend to be *casual* users. A casual user is someone who occasionally puts together a report, a newsletter, or a presentation. Casual users are often willing to live with a limited set of features in exchange for ease of use, therefore software intended for casual use must be easy to learn. It must also be relatively inexpensive, and run on more or less "standard" desktop computers. The most popular applications for general business users are word processors, spreadsheet programs, database management systems, and project planning tools.

Today, most of these applications have graphical interfaces. Pioneered in the 1970s by Xerox at the Palo Alto Research Center, the graphical user interface has since become the standard way for users to interact with their computers. As a result, different applications can share common conventions for performing various tasks, thereby reducing the learning curve for new applications. In addition, the popularity of the graphical user interface has been a driving force behind an increase in the graphics capabilities of the hardware. In both PCs and workstations, the graphics support has increased tremendously both in terms of resolution (resulting in crisper, clearer images) and in terms of the number of colors available (resulting in richer and more natural looking images). This has, in turn, enabled *new* software applications to be developed that exploit the improved graphics capabilities. These include presentation graphics packages as well as draw and paint programs.

It is in the area of presentation graphics that multimedia had its initial impact for casual business users. Presentation packages such as Microsoft's Powerpoint are designed primarily to produce slides or transparencies for display using a projector. However, with the improved graphics capabilities, the computer itself can be used as a presentation device.

There are a number of advantages and disadvantages to this. An obvious disadvantage is the need to have a computer available in order to present the material. Among the advantages is the ability to make presentations more dynamic—for example, transition effects (e.g., wipes or fades) can be added between slides to make the presentations more visually stimulating.

Presentation software has benefited from the advent of inexpensive audio hardware. Sound can add a great deal of impact to a presentation all by itself, and is of course a key element in software systems such as Microsoft's Video for Windows and Apple Computer's Quicktime. Both Video for Windows and Quicktime provide a standard level of support for digital audio and video data on desktop computers. This makes it easier for developers to incorporate audio and video into their existing applications. Newer applications, such as Compel from Asymetrix, have been designed from the ground up to support multimedia data types, and include features such as animated graphics and text objects, and cues for triggering sound or video clips, in addition to the more traditional transition effects.

Using the computer as a presentation device opens up the possibility of integrating audio and video data into other applications besides presentation graphics. Word processors, which are typically used to produce paper documents, can be used to produce documents that are intended to be read on-line (i.e., read on the computer screen, rather than printed on paper). It then becomes possible to add audio and video to text, graphics, and images to create multimedia documents. Documents that contain several data types are often called compound documents, and are generally created using a collection of integrated software tools. Individual data elements may be created (or captured) and edited using data-specific tools and later combined into an integrated document.

A compound document viewer application may present a single unified interface to the user, but call upon data-specific software components when rendering the data. In other words, the various software components must interoperate. This is an important feature of multimedia applications.

Casual users (like all users) work in a preexisting environment, with familiar software tools and preexisting materials. Multimedia support is generally an incremental, not an essential set of features for most general business users. If these features are to be widely used, they must, for the most part, be added to the existing environment in an unobtrusive and seamless way.

Specialized Business User

Specialized business users are also a diverse group. These individuals bring specialized knowledge or skills to the performance of their tasks, and are anything but casual users of the technologies they employ in the process. Examples of specialized users of multimedia systems are: publishers, graphics designers and artists, and product design engineers. These professionals generally use a collection of software applications to perform various interrelated tasks, and quite often work in coordination with other specialists in teams. Specialized users, like general business users, are interested in interoperability between applications—in many cases demanding even tighter integration between the various systems. In addition, they are often concerned with the flow of the work process itself, particularly when they are working in teams.

Word processors, desktop publishing software, and digital image processing systems are available from a wide variety of vendors. These systems have reached such a level of sophistication and ease of use that they are almost taken for granted, and have forever altered the way in which published works are produced. Instead of typesetting text manually, it is created using a word processor with type fonts, point sizes, and other formatting parameters selected from a menu. Rather than resize and crop photographic images, photographs are scanned into the computer (i.e., digitized) where they can be more easily manipulated. The layout of the page can then be done entirely under control of the computer, with the computer's monitor providing immediate visual feedback. The data can then be sent to a printing device where a high-resolution version of the final product can be generated. These software systems have replaced other methods and processes used in the production of printed materials, and have gone on to provide a higher level of integration between the various steps involved in production. Creation of individual components such as text, graphics, and images can be done using specialized application software, and then later combined. Small corrections or changes do not require major steps to be repeated, since the digital information can be more easily modified. This work can be done by a single individual, but more often is performed by a team of specialists. Using digital technologies, these teams can pass data and work-in-progress around to one another digitally over a network.

Graphics artists have also embraced digital technologies. In a survey conducted in the summer of 1988 by *Communications Arts* magazine, 70% of respondents said that they owned computers. Of these, 68% listed their primary business as graphics design. Most of these said they used computers for production, as well as in the design process. Among the software systems listed most often were the ubiquitous word processors such as Microsoft Word and MacWrite, desktop publishing systems such as Aldus PageMaker and Ventura Publisher, and several graphics systems including Adobe Illustrator, Lotus Freehand, and MacDraw.[11]

Digital manipulation of photographic images is becoming an increasingly common part of the graphics design process. Once digitized, these images can be altered or enhanced, resized, cropped, rotated, and ultimately combined with other elements such as text and graphics. Sophisticated digital imaging software (such as Adobe Photoshop) allows the user to adjust color values, create composite images, and perform filtering operations to lighten or darken, soften or sharpen, or otherwise alter the image.

The growth of these specialized software systems, and their subsequent integration into a tool suite for producing printed media, has now begun to impact the workflow of

---

11. Wendy Richmond, *Design & Technology*, Van Nostrand Reinhold, 1990, pp. 126–133.

the organizations using them. This has led to the emergence of integrated software systems designed specifically with workflow in mind, such as the Quark Publishing system, and open application architectures such as the Publication Administrator System (from North American Publishing Systems). An open application architecture system is one which allows products from different vendors to be integrated into an overall system.

It is clear that digital technologies have had a considerable impact on the way graphics design, image processing, and publishing is done. It has also had an additional effect by making it possible to publish information electronically. CD-ROM technologies have played an important role in this. The information itself had already come under the control of the computer; with the introduction of the CD-ROM, it became possible to distribute the material in digital form as well. Electronic publishing has evolved along the same lines as general business presentation systems. Once the computer is used as a presentation device, it is possible to add dynamic elements like audio and video that were impossible with static media such as print. It is here that the domain of the application developer and the publisher begin to converge.

A traditional application developer creates a software system to automate a process or manipulate or manage data, but does not typically include the data themselves (except as samples). In these cases, the user has his/her own data, and is simply looking for a better way of dealing with them. A media producer, like a publisher or a film maker, creates data (sometimes called *content*). When an application includes its own data it is referred to as a *title*. Electronic encyclopedias such as Microsoft's Encarta and Compton's Multimedia Encyclopedia are good examples of titles. In a title, the content is what the user really wants. The application is there mainly to allow the user convenient access to the content.

It is possible to create a title using a slide-based presentation package, but in general these applications are too limited in their capabilities for this purpose. For users needing more power and flexibility, there are hyperlink or script-based authoring systems. Hyperlink authoring software provides functions for linking multimedia data to on-screen "hot-spots" which the user can click on with a mouse or activate with a touch screen. They can also link various data elements together so that a user can easily find related material. This kind of authoring is used by developers of on-line help systems, and computer based training applications, as well as multimedia catalogs and other kinds of multimedia documents.

Script-based tools provide the user with a special-purpose programming language that can be used to produce sophisticated titles and applications. And of course for the hard-core multimedia developer there are traditional development environments and languages such as C or C++. A growing number of authoring systems provide an integrated set of tools which allow the user/developer to import multimedia data from a variety of sources, place them as a sequence of events along a timeline, add effects and links, and perhaps even add in specialized modules written in a programming language or script.

Another kind of specialized business user is the design engineer now using digital systems to aid in the design of manufactured products. Designers and engineers have probably been using drawings and models in their work since the beginning of time. It should not be a surprise to find them drawing and modeling on computers. The initial systems were simply automated 2D drafting tools, but these have since evolved into a complex set of software tools referred to as Computer-Aided Design (CAD) software. CAD software is used in the design of every kind of product from microchips to office buildings. CAD is used not only for creating the "blueprints" associated with the design of a product, but also for analyzing various aspects of the design. For example, CAD software for designing integrated circuits might include tools for simulating the circuits' behavior, or performing a check on the manufacturability of the chip against the "design rules" for a given chip fabrication process.

A large number of products can be designed using CAD tools that model surfaces using 2D and simple 3D geometry. However, for stylized products, and those that need to take advantage of the laws of fluid dynamics (such as aircraft and automobiles) more sophisticated tools are required. These products typically have smooth freeform surfaces that must be modeled, requiring the CAD software to support mathematical modeling techniques as well as drawing techniques.[12] As it turns out, these features are also useful for creating "realistic" models for other purposes, such as 3D animation, of which more will be said later.

A good example of the use of computer-aided design is the recent project undertaken by Sweden's Volvo Corporation, which used a modeling software system called Alias Studio to design an extremely low-emitting concept car. The car needed to be aerodynamic and still be able to comfortably seat the family. Their design process began with several manual sketches which were later scanned into the computer. From there the 3D modeling software was used to turn the sketches into a 3D model of the car. Both the interior and exterior of the car were modeled in software.[13]

Architects also make heavy use of CAD, and are beginning to make use of multimedia software as well. In addition to traditional plans and drawings, architects are beginning to add photographs, animations, and even audio and video to help their clients visualize the final product. Some architects are even beginning to use virtual reality software in their work, enabling clients to conduct electronic "walk-throughs" so that any changes can be made before construction begins.

As we saw with publishing and graphics design users, CAD users have also begun to change the workflow to take better advantage of the automated design process. These changes began with the simple exchange of design docu-

12. L. Stephen Wolfe, "Freeform Surfaces," *Computer Graphics World*, July 1993, pp. 59–64.
13. Diana Phillips Mahoney, "A Concept for a Caring Car," *Computer Graphics World*, May 1993, pp. 49–55.

ments between the designers themselves. Next, the need to exchange data between the various software tools used in the design process gave rise to integrated software systems and open architecture systems that automate the process of data interchange. Unlike desktop publishing and graphics design, CAD software must also be integrated with other software systems for performing such tasks as design rule checking, airflow analysis, or operational simulation. Finally, in some industries, CAD software is being integrated with Computer-Aided Manufacturing (CAM) software to create completely automated design and manufacturing systems.

Groupware User

Groupware is a class of software designed to support the activities of individuals working as a group. The most common and obvious groupware application is electronic mail. The importance of groupware applications as well as of making current applications group-*aware* will continue to increase as the percentage of computers installed in networked environments increases. In addition, there is currently a surge of activity among hardware manufacturers to integrate telecommunications capabilities into desktop computer systems. The emergence of groupware applications is an important trend for multimedia. They deal with audio and video as data types in a distributed processing and distributed data environment. In this application, the temporal requirements are more stringent (as with teleconferencing for example) and getting the data across networks on demand is more important than the quality.

Groupware applications are designed to facilitate human-to-human interaction. The goal of these systems is literally to transcend the limitations of time and space, ultimately providing location transparency for all shared resources (including information) and people. Groupware aids in the communication and coordination of activities of people working in groups and provides them with an interface to a shared environment. These applications can be categorized according to the data types they support (e.g., text, graphics, audio, video), and whether they enable synchronous or asynchronous communication. It is also common to place groupware applications in a 2 × 2 matrix according to their temporal and spatial context:

	Synchronous (Same Time)	Asynchronous (Different Time)
Distributed (Different Place)	Audio Conferencing Video Conferencing Shared Workspace	Electronic Mail Computer Conferencing Voice Mail
Co-Located (Same Place)	Electronic Meeting Room	Document/Version Control Media Library

The most familiar example of groupware is the computer-based message system known as electronic mail, which supports the asynchronous exchange of textual messages between groups of users. Computer conferencing is simply a structured form of electronic mail, where several individuals participate in an asynchronous electronic discussion related to a specific topic. The discussion is often goal directed, and participants are generally directed to submit ideas, critique the ideas of others, and indicate areas of agreement. Computer conferencing is also commonly used as a forum for ongoing discussion about a topic of interest to the participants. Asynchronous shared workspace applications include document/version control systems, distributed database systems, and other multi-user systems. For the most part these systems work by providing locks for all or part of the data to prevent multiple users from altering it at the same time. Asynchronous shared workspace applications enable users to simultaneously see and manipulate shared data. These applications include shared text and graphics editors, and electronic "white board" systems.

Voice mail systems allow users to send, receive, and manage voice mail messages. Digital voice mail systems have recently become available and are now being integrated into desktop computer systems. Audio-conferencing applications enable two or more individuals to participate in a real-time peer-to-peer audio conversation. Video-conferencing applications enable two or more individuals to participate in a real-time audio-visual conversation, or enable an audio-visual presentation to be broadcast to multiple client sites.

Applications that support media spaces are similar to synchronous shared workspaces except that they may incorporate any or all forms of media. When the application supports distributed participants it is classified as a *desktop conferencing* system. Such a system uses the PC as the conference interface, allowing multiple groupware applications to run simultaneously. Sophisticated systems will support multiple video windows per station, allowing users to create, edit, and view information in a variety of forms while participating in a video conferencing session.

Media Producers

Traditional techniques of creating animation involve hand drawing individual frames of a sequence and then photographing each one, a frame at a time, until the entire sequence has been recorded. The sequence is then played back at a high speed, thereby producing the illusion of motion. When you consider that a lot of the image stays the same from frame to frame, and that film speed is 24 frames per second, it is easy to see that the amount of work to produce animations manually is staggering. In addition, the degree of realism that can be achieved with traditional techniques is somewhat limited. High-end computer animation systems can produce results that are good enough to fool (even impress) feature film audiences.

In the film *Jurassic Park*, a collection of 75 Silicon Graphics workstations running a combination of commercial software and custom programs was used in the creation of animated dinosaurs that were then combined with live action footage. The commercial software used by the animators included *Alias* for constructing wireframe models of the dinosaurs, *Softimage* for animating the models, *Colorburst* (now called MATADOR paint, from Parallax Graphics Systems) for painting in the textures, and *Renderman* (from Pixar) for lighting and rendering. The custom software was developed at Industrial Light and Magic to link the commercial software to extend their functionality.[14]

It is possible to achieve "broadcast quality" results with much less cost and effort (albeit at the expense of the super-realism achieved by the likes of ILM) using desktop-based systems such as *Crystal TOPAS Professional* (for the PC) and *MacTOPAS* (for the Macintosh) from CrystalGraphics, *3D Studio* (for the PC) from Autodesk, *LightWave* (for the Amiga) from NewTek, *Caligari Broadcast* (for the Amiga) from Octree Software, and *Electric Image Animation System* (for the Macintosh) from Electric Image. Such systems combine the features of 2D and 3D modeling, color and texture mapping, animation, and rendering into a single integrated system.[15]

Digital systems are also being used for video editing. A number of digital-editing software systems exist for desktop computers, and are generically called Desktop Video (DTV) Systems. DTV editing systems attempt to replace analog editing systems and therefore are classified in the same categories based on the kind of editing they can perform, which was originally based on how many tape decks (VTRs) they were designed to control at once. In a Cuts-Only Edit System, the computer controls two VTRs, and enables the user to edit different segments of video together from the first deck to the second deck, without the addition of graphics or transition effects. This could be used to remove the commercials from the movie you just recorded, for example. In an A/X Roll Edit System, the computer uses a video graphics adapter and a single VTR, which enables the user to cut between computer-generated materials and video. An A/B Roll Edit System uses the computer, three VTRs, and special effects hardware, enabling the user to create transitions (fades, dissolves, wipes, etc.) between different segments of video.

Of course, there are now all digital, disk-based editing systems which store the video to be edited in binary form on

---

14. Barbara Robertson, "Dinosaur Magic," *Computer Graphics World*, September 1993. pp. 44–52.
15. Steven Blaize, "Traveling Exhibition: Works of Art That Move," *Desktop Video Works*, February/March 1993, pp. 84–91.

large hard disks where they can be manipulated directly without having to seek locations on the tape. These systems generally allow the user to capture, edit, and view the video under the control of the computer and within the digital domain. Some of these systems also allow the addition of titles and graphics, or special effects—particular transitions such as wipes and dissolves. The final product can then be mastered in a variety of either digital or analog formats. These systems vary widely in price and capability, with the high-end systems often being sold as part of a complete hardware/software system. These include Avid's Media Suite Pro (for the Macintosh), Matrox Studio (for the PC), and the Amiga-based Video Toaster. For more casual users, or DTV professionals targeting distribution of CD-ROM titles (as opposed to broadcast video) there are lower cost software systems such as Adobe Première, and DiVA VideoShop 1.0 (MAC).

Media Consumer

By far the largest potential user base for multimedia software is the mass market consumer. Cartridge-based software for entertainment systems from companies like Nintendo and Sega have been available for some time and have sold huge numbers of systems. CD-ROM based versions of these game machines have also appeared.

In addition, the number of general purpose desktop computers equipped to support multimedia is also growing. Growth in the number of users in this category has been slow due to a lack of standards for publishers in addition to a relatively small installed base of hardware systems to support them. Nevertheless, more and more systems are being shipped with CD-ROM drives and audio support as standard equipment, and standard multimedia configurations such as MPC and MPC II have emerged to at least provide a standard platform. Most of the software targeted at the mass market are either interactive games, such as 7th Guest and King's Quest, or reference titles like the multimedia encyclopedias already mentioned.

The Multimedia PC Marketing Council
Level 2 Specification

The Multimedia PC Marketing Council, Inc. (a group of hardware and software vendors) issued their a second-level multimedia computer specification in May of 1993 to encourage the adoption of enhanced multimedia capabilities. This specification is a backwardly compatible superset of the MPC Level 1 Specification, which continues in effect. This Specification defines the minimum system functionality for Level 2 compliance, but is not intended as a recommendation for a particular system configuration. The council is considering plans for an MPC Level 3 during mid-1995, but has not yet made any formal announcements.

The MPC hardware certification program is being extended to CD-ROM drives and sound cards. The Council has also announced the adoption of an analog audio cable standard for these multimedia components.

Hardware Specifications

CPU—*Minimum requirement:* 25 Mhz 486SX (or compatible) microprocessor.

RAM—*Minimum requirement:* 4 megabytes of RAM (8 megabytes recommended).

Magnetic Storage—*Requirement:* 3.5" high density (1.44 MB) floppy disk drive.

Requirement: 160 MB or larger hard drive.

Optical Storage—*Requirements:* CD-ROM drive capable of sustained 300 KB/sec transfer rate;

no more than 40% of the CPU bandwidth may be consumed when maintaining a sustained transfer rate of 150 KB/sec.;

average seek time of 400 milliseconds or less 10,000 hours MTBF;

CD-ROM XA-ready (mode 1 capable, mode 2 form 1 capable, mode 2 form 2 capable);

multisession capable;

MSCDEX 2.2 driver or equivalent that implements the extended audio APIs;

subchannel Q support (P, R-W optional).

Recommendations: At 300 KB/sec sustained transfer rate, it is *recommended* that no more than 60% of the CPU bandwidth be consumed.

It is *recommended* that the CPU utilization requirement and recommendation be achieved for read block sizes no less than 16 K and lead time of no more than is required to load the CD-ROM buffer with 1 read block of data.

It is *recommended* that the drive have on-board buffers of 64 KB and implement read-ahead buffering.

Audio

Requirements: CD-ROM drive with CD-DA (Red Book) outputs and volume control.

16-bit Digital-to-Analog Converter (DAC) with linear PCM sampling; DMA or FIFO buffered transfer capability with interrupt on buffer empty; 44.1, 22.05, and 11.025 kHz sample rate mandatory; stereo channels; no more than 10% of the CPU bandwidth required to output 22.05 and 11.025 kHz; it is *recommended* that no more than 15% of the CPU bandwidth be required to output 44.1 kHz.

16-bit Analog-to-Digital Converter (ADC) with linear PCM sampling; 44.1, 22.05, and 11.025 kHz sample rate *mandatory*; DMA or FIFO buffered transfer capability with interrupt on buffer full; microphone input.

Internal synthesizer capabilities with multivoice, multitimbral capacity, 6 simultaneous melody voices, plus 2 simultaneous percussive voices.

Internal mixing capabilities to combine input from three (*recommended* four) sources and present the output as a stereo, line-level audio signal at the back panel. The four sources are CD Red Book, synthesizer, DAC (waveform), and (*recommended but not required*) an auxiliary input source. Each input must have at least a 3-bit volume control (8 steps) with a logarithmic taper. (4-bit or greater volume control is strongly recommended.) If all sources are sourced with –10 dB (consumer line level: 1 milliwatt into 600 ohms = 0 dB) without attenuation, the mixer will not clip and will output between 0 dB and +3 dB. Individual audio source and master digital volume-control registers and extra line-level audio sources are *highly recommended*.

CD-ROM XA audio capability is *recommended*.

Support for the IMA-adopted ADPCM software algorithm is *recommended*.

(Guidelines for synthesizer implementation available on request.)

Video

Requirement: Color monitor with display resolution of 640 × 80 with 65,536 (64 K) colors.

The *recommended* performance goal for VGA+ adapters is to be able to blit 1, 4, and 8 bit-per-pixel DIBs (device independent bitmaps) at 1.2 megapixels/second, given 40% of the CPU. This recommendation applies to run-length encoded images and nonencoded images. The recommended performance is needed to fully support demanding multimedia applications, including the delivery of video with 320 × 240 resolution at 15 frames/second and 256 colors.

User Input

Requirement: Standard 101 key IBM-style keyboard with standard DIN connector, or keyboard that delivers identical functionality utilizing key-combinations.

Requirement: Two-button mouse with bus or serial connector, with at least one additional communication port remaining free.

I/O

Requirement: Standard 9-pin or 25-pin asynchronous serial port, programmable up to 9600 baud, switchable interrupt channel.

Requirement: Standard 25-pin bidirectional parallel port with interrupt capability.

Requirement: 1 MIDI port with In, Out, and Thru; must have interrupt support for input and FIFO transfer.

Requirement: IBM-style analog or digital joystick port.

System Software

Multimedia PC system software must offer binary compatibility with Windows 3.0 plus Multimedia Extensions, or WindowsL 3.1.

Minimum Full System Configuration

A full Multimedia PC Level 2 system requires the following elements and components, all of which must meet the full functional specifications outlined above:

CPU	25 Mhz 486SX or compatible microprocessor
RAM	4 megabytes of RAM (8 megabytes recommended)
Magnetic Storage	Floppy drive, hard drive (160 MB minimum)
Optical Storage	CD-ROM doublespeed with CD-DA outputs, XA ready, multisession capable
Audio	16-bit DAC, 16-bit ADC, music synthesizer, on-board analog audio mixing
Video	Display resolution of at least 640×480 with 65,536 (64 K) colors
Input	101 key keyboard (or functional equivalent), two-button mouse
I/O	Serial port, parallel port, MIDI I/O port, joystick port
System Software	Binary compatibility with Windows 3.0 plus Multimedia Extensions, or Windows 3.1

Minimum Upgrade Kit Configuration

A Multimedia PC Level 2 Upgrade Kit requires the following elements and components, all of which must meet the full functional specifications outlined above:

Optical Storage	CD-ROM doublespeed with CD-DA outputs, XA ready, multisession capable

Audio	16-bit DAC, 16-bit ADC, music synthesizer, on-board analog audio mixing
I/O	MIDI I/O port, joystick port

(Providing system software with Upgrade Kits is optional.)

CD-ROM/Sound Card Audio Cable Standard for MPC Components

The following cable standards apply only to MPC components (CD-ROM drives or sound cards sold separately). Full systems and upgrade kits are not required to observe the following specification:

A Multimedia PC CD-ROM drive component must include a minimum 24-inch cable to connect the drive's analog audio output connector to an MPC sound card's analog audio input connector. The cable's open sound card connector must be a female 4-pin Molex 70066–G, 70400–G, or 70430–G connector with 2.54 mm pitch, or the equivalent, with the following pin assignments: pin 1 – left signal, pin 2 – ground, pin 3 – ground, pin 4 – right signal.

A Multimedia PC sound card component must be capable of mating with the CD-ROM audio cable by having a 2.54 mm pitch Molex 70553 male connector on the card (or the equivalent), or by including a short patch cable. The patch cable must plug into the nonstandard sound card connector and have an open male connector (Molex 70107–A, or the equivalent) for attaching to the CD-ROM cable female connector. The pin assignments on the sound card connectors must be complementary to the CD-ROM audio cable connector.

The CD-ROM/Sound Card Audio Cable Standard has also been added to the MPC Level 1 Specification. For a complete copy of the Level 1 Specification, please contact the Council.

Also contact the Council for a Summary of MPC Certification Mark Hardware Licensing Requirements.

Glenn Ochsenreiter
Managing Director
Multimedia PC Marketing Council, Inc.
1730 M Street, NW, Ste. 707
Washington, DC 20036
202-331-0494

Core Multimedia Software Technologies

Fundamental Operating System Issues

Media types come in two flavors, *discrete* and *continuous*. Discrete media types are time independent (i.e., their presentation is not necessarily a function of time). Text, graphics, and still images are all examples of discrete media types. Continuous media types, such as audio and video, are time-dependent in that their correct presentation is a function of time. Continuous media types and are also referred to as isochronous data.[16] The processing of all digital data, regardless of their temporal aspect, involves a common set of tasks such as creating, modifying, storing, retrieving, transmitting, and rendering. There are a number of services typically provided by the operating system which aid the application in the performance of these tasks. The fundamental issues facing an operating system designer who is attempting to provide support for multimedia are how to modify and extend these services to support continuous media.

Continuous media within the digital domain are not actually continuous, but is made up of a series of samples. It is the requirements placed on the processing of these samples with respect to time that makes the media continuous. In other words, each sample must be delivered and processed within a specified amount of time that corresponds to the sample rate. This in turn places stringent requirements on the systems resources in terms of: throughput (the system must be able to continuously deliver samples in time to be processed); and processing power (the system must be able to process each sample in time).

The hardware necessary for processing digital audio and video data has only recently become available for desktop PCs and workstations. These most commonly take the form of optional add-in modules designed to plug into existing systems; however, as these features become more popular, they will become more tightly integrated into the overall sys-

---

16. Steinmetz, Ralf, "Synchronization Properties in Multimedia Systems," *IEEE JSAC*, Vol. 8, No. 3, April 1990.

tem. In this way, the data-processing computer evolves into a *video computer*[17] or a *telecomputer.*[18] This evolution, which is currently under way, will result in new "standard hardware" configurations (e.g., the Multimedia PC; see pages 204–208 for the MPC Level 2 Specification) where support for audio and video is as common as printer or keyboard support is today. Integrating high-bandwidth continuous media such as video will also require additional capacity on traditional systems resources such as the CPU, memory, secondary storage, the system bus, and the network. (See Chapter 6.)

Of course there are always cost considerations associated with hardware resources or the solution would be simply to provide more MIPs and megabytes. So we proceed by assuming that all resources are limited, and that to effectively utilize them, they must be properly managed. Proper management of resources involves making sure that (whenever possible) resources are available where and when they are needed. Therefore when we speak of *resource management,* we mean not only the allocation of resources, but also the efficient *scheduling* of their use. In addition, multimedia applications typically want to make use of a collection of resources in a coordinated way. This introduces another issue that must be addressed that is related to scheduling, namely *synchronization.* The quintessential example of synchronization is (as mentioned earlier) the simultaneous playback of related audio and video material, commonly known as lip-sync.

Resource Management

Resource management refers to an operating system service that coordinates access to system resources by applications. An application will typically make a resource request of the

---

17. Smith, Alvy Ray, "The Video Computer: Image Computing in the Studio."
18. Clark, Jim, "A TeleComputer," *Computer Graphics*, Vol. 26, No. 2, July 1992.

resource manager, which checks to make sure that there is sufficient capacity available for the requested resource and either grants the request (by allocating the required access to the resource), or denies the request (in the event that sufficient capacity does not exist). Sophisticated resource management schemes that provide for the dynamic reallocation of resources depending on the changing needs of the application(s) are also possible. A good statement of the requirements for a multimedia resource management system is given by Hanko et al.: "What is called for here is a general systems solution to the problem of managing time-critical activities. It is not sufficient for the system to simply "get out of the way." Rather, it is the system's responsibility to actively manage its resources in such a fashion as to, whenever possible, permit the applications' time-constraints to be met."[19]

Multimedia applications typically combine a variety of media types into a single "presentation" and may need to share the physical resources of the system among various media-processing tasks. In addition, the application may need to share resources with other applications (even with other multimedia applications). Making it possible to share resources among applications results in better overall resource utilization as long as the total requirements of the application(s) do not exceed the capacity of the resource. In practice, the capacity of resources is always limited, and occasionally, an application (or set of applications) will have total requirements that exceed the capacity of a given resource. In addition, the processing of continuous media data requires that a collection of resources be allocated and coordinated to insure that the processing requirements are met end-to-end. This means that virtually all the systems resources should be under the control of the resource manager.

---

19. Hanko, James G., et. al., "Workstation Support for Time-Critical Applications," *Network and Operating System Support for Digital Audio and Video: Second International Workshop Proceedings*, Springer-Verlag, November 1991.

When the continuous media system is also a distributed system, resource management becomes more complex. A distributed application will make use of resources on more than one physical machine, and the activities of these resources must be coordinated across machine boundaries. Managing resources in a distributed environment requires a form of resource management in which local resource managers can coordinate their activities with one another on behalf of a distributed application. An example of this kind of resource management is described by David P. Anderson of the University of California at Berkeley: "The task of an IDCM [integrated digital continuous media] application is simplified if it can be "guaranteed" that the system will handle CM data with the necessary performance levels for the duration of the execution. In order to make such a guarantee, the shared components, such as the CPU, file system, and network, must support "reservations" in which the client specifies its workload and the component provides a performance guarantee. Furthermore, to provide end-to-end guarantees it may be necessary to use a meta-scheduler to coordinate components. A meta-scheduler is a distributed software layer that "reserves" components on behalf of client applications; it is not involved in the actual usage of the components."[20]

Real-Time Scheduling

Common desktop and workstation operating systems are designed to support a wide variety of tasks. These tasks include a few time-critical tasks, sporadically mixed in with a larger number of non–time-critical tasks. One of the components of the operating system (called the scheduler) is responsible for deciding which task should be run next, and which tasks can wait their turn. Special time-critical tasks can interrupt the processing of a currently running task in order for the time-critical task to run, after which the previously

---

20. Anderson, David P., "Meta-Scheduling for Distributed Continuous Media," Computer Science Division, EECS Department, University of California at Berkeley, October 4, 1990.

running task can resume processing. Even for normal tasks, some may be more important than others, for example the task of sending data to a printer may be less important than sending data to the display. In order for the scheduler to make intelligent decisions about which task to run next, they are often assigned priorities. Tasks with higher priorities are allowed to run before lower priority tasks.

Most operating systems have schedulers that are designed to share the computer's processing resources more or less fairly among the various tasks, and may even increase a task's priority if it has been waiting an unusually long time for a chance to run. If a task is not allowed to run for a while, the system may appear sluggish to the user (sometimes resulting in an annoying little hour-glass on the display where the cursor ought to be), but this is generally not considered a system failure. In a real-time system, time-critical tasks are the norm rather than the exception to the rule. In these systems, each task is often assigned a priority that is selected to insure that it is processed in time. If a real-time task is not allowed to run, and cannot be completed on time, it is considered a system failure.

Early digital video systems simply took over all the resources of the computer and essentially replaced the general purpose scheduler with their own real-time scheduler. This was the approach taken by the original Intel DVI system from RCA (later Intel). This system was known as the Audio Video Support System (AVSS).[21] It managed the tasks associated with processing digital audio and video using a real-time executive module called RTX. AVSS ran on PCs running either MS-DOS or PC-DOS, but the RTX module essentially took over the scheduling of tasks from the DOS scheduler while AVSS was active. This was done to insure that the time-critical tasks associated with audio-video processing were given preferential treatment over any other tasks. The obvious result of this technique is that no other applications can be running while AVSS is active. Since DOS

21. Arch C. Luther, *Digital Video in the PC Environment*, McGraw-Hill, 1991, pp. 214–223.

is a single task operating system, there is only one application running at a time anyway, so this approach did not severely limit the normal operations of the computer.

Another approach is to develop a special purpose embedded real-time operating system. This is the approach used in consumer-oriented multimedia players such as Compact Disk-Interactive (CD-I), a system developed by Philips, Sony, and Microware corporations. A CD-I player connects to standard home television and stereo equipment.[22] The CD-I player hardware includes a decoder chip, a block of DRAM, a video-keying subsystem, and a dedicated CPU, and a ROM containing a real-time operating system called CD-RTOS (Compact Disk Real-Time Operating System). A system controller inside the decoder chip is responsible for satisfying the real-time requirements of the playback stream based on timing data extracted from the encoded bitstream.[23]

Neither of these approaches is satisfactory for use on the current generation of multitasking operating systems used in desktop PCs and workstations. Unlike DOS, these systems allow more than one application to share the CPU, and unlike the multimedia players, are intended to be general-purpose machines. New strategies have been (and continue to be) developed to enable the time-critical tasks associated with temporal data processing to coexist in a standard environment with more traditional tasks associated with spreadsheets and word processors.

Synchronization

There are two types of continuous media synchronization of concern to us: *intrastream synchronization* (the synchronization of events that occur during the processing of a stream); and *interstream synchronization* (the synchroniza-

---

22. Frenkel, K. A. "The Next Generation of Interactive Technologies," *Communications of the ACM*, Vol. 32, No. 7, July 1989, pp. 872–881.
23. Sijstermans, F., and J. van der Meer, "CD-I Full Motion Video Encoding on a Parallel Computer," *Communications of the ACM*, Vol. 34, No. 4, April 1989, pp. 81–91.

tion of one stream with another). The processing of a continuous media stream from its source to its destination involves a number of steps which must be performed in the proper order, and within a specified time frame in order to insure that the result appears continuous to the human observer. Each of these processing steps takes some time to perform and thus introduces a delay. In addition, the amount of time it takes to perform each step may vary, which introduces jitter. Intrastream synchronization is the coordination of these processing steps so that the cumulative delay and jitter can be accounted for, and the illusion of continuity preserved. When two or more continuous media streams need to be coordinated in time, as in the lip-sync case, another temporal anomaly is introduced, namely skew. Interstream synchronization is the coordination of the processing of two or more continuous media streams so that the skew between them is bounded in such as way as to preserve the illusion of simultaneity.

Evolution of Key Operating Environments

Graphics Support

Operating systems have been evolving for several years to better support graphical data, due to the increasing popularity of graphical user interfaces. Displays for desktop computers have moved from character-based monochrome, through several "standards" such as CGA (Color Graphics Array), EGA (Extended Graphics Array), VGA (Video Graphics Adapter), and SVGA (Super Video Graphics Adapter). Each new standard specifies higher resolution and more colors. At the same time, workstation graphics has steadily improved as well, supporting very high resolutions and full color. Combined with improvements in processing speed and the reduction in cost of memory and storage, this has allowed for the digital manipulation of nontemporal data types—which has enabled such applications as desktop publishing and image processing.

QuickTime™

Apple QuickTime[24] is a set of extensions to the Macintosh operating system which provide support for "dynamic data types" such as audio, video, and animation. Quicktime takes a number of concepts from the manipulation of spatial objects (e.g., stretching and shrinking, and cut and paste), and applies them to temporal objects. Stretching or shrinking a temporal sequence increases or decreases the play time of the sequence. Cutting and pasting allows a sequence to be removed, and/or imbedded into another sequence, and so on.

The Quicktime architecture includes the specification of a special kind of compound media file format called a "movie file" which enables multiple tracks of media to be treated as a single entity. Each track can have its own media type and sample rate, and each track can be offset from the time origin of the movie file as a whole. QuickTime movie files store pointers to the actual media elements, allowing editing operations to be performed without copying the bulk of the data. This same feature also allows multiple movie files to have different edit lists indexing into a common set of media elements, thereby reducing overall storage requirements.

The QuickTime architecture also includes a set of "toolboxes" for installing and managing image compressors, digitizers, interface elements, and other components of the system, and for controlling and synchronizing temporal objects during playback and editing. For the latter, Quick-Time maintains a "global clock," which drives all time-based events and a common controller that allows tracks with different sampling rates to be synchronized. This common controller also has a standard user interface component so that applications can provide a consistent interface for controlling temporal media.

---

24. Hoffert, Eric, et. al., QuickTime™: An Extensible Standard for Digital Multimedia, *CompCon92 Proceedings*, IEEE Computer Society Press, February 1992, pp. 15–20.

Windows Multimedia

Microsoft Windows supports multimedia and digital video computing through a collection of system elements, including the Audio Video Interleaved (AVI) file format, the Media Control Interface (MCI), Video for Windows (VFW), and Object Linking and Embedding (OLE) technology.[25]

Audio Video Interleaved (AVI) is an extension of RIFF (Resource Interchange File Format) designed to store digital audio and video information. This format is also designed to be cross-platform compatible, allowing content on Windows-based systems to play on other operating systems. Files that have the '.AVI' extension are AVI files, and the audio and video information contained in them can be accessed, manipulated, and preserved by the full range of Windows-compatible hardware and application software.

The Media Command Interface (MCI) allows all Windows-compatible application software to control a variety of multimedia devices, including CD-ROM drives, audio, and animation players. The player "engine" for controlling the rendering of AVI files is an MCI driver, based on the Digital Video command set (DV-MCI) developed jointly by Microsoft, Intel, IBM, and others.

As with the Windows operating system family itself, a major advantage of the Video for Windows architecture is its scalability. Video can scale in multiple dimensions, including resolution, frame rate, and color depth. Video for Windows automatically takes advantage of all the capabilities of the system it is running on, to scale in any or all of these dimensions. Users do not have to worry about which flavor of video-compatible PC they must use to play a particular video sequence. Because all Windows-based products support OLE, MCI, AVI, and the other architectural elements of video and multimedia, users can play back digital video sequences on any of these platforms, and use them interchangeably.

---

25. Green, James L., "Multi-Platform Support for Digital Media Opens New Markets," *Computer Technology Review*, February 1993, pp. 26–28.

The object linking and embedding (OLE) technology lets users insert multimedia elements, including digital video, into software programs, and is therefore automatically compatible with more than 150 OLE-aware desktop applications, including Microsoft Word, Microsoft Excel, and applications from third-party vendors.

The "Architected" Environment

Over the years, software engineers and researchers have developed a variety of "systems" or "methodologies" for determining the design and construction of software systems. Some of these were even useful. All of them reflect the kinds of problems the methodologists were studying at the time. From an historical perspective, each time the level of system complexity went up, new methodologies were invented in an attempt to cope with it. The latest of these design methodologies is called object-oriented design. Nearly all the multimedia systems under development today have object-oriented designs. Therefore, it is important to understand something about object-orientation before proceeding.

First of all, designing and constructing object-oriented software benefits mainly the software developer. Object-orientation allows the software developer to design and build systems out of a collection of self-contained, but interactive components. These components, if designed correctly, can provide a number of benefits to the developer. Since components are self-contained (this is called *encapsulation*), one component can be replaced by another without affecting the rest of the system. This makes maintenance of the system easier. Components can provide services to many different applications (this is referred to as reuse). This has the effect of reducing the total amount of code that needs to be developed. Objects are typically arranged in hierarchies that represent a parent/child relationship. This allows specialized (child) objects to be easily defined in terms of more general (parent) objects (this is referred to as the principle of inheritance).

Of course, users ultimately benefit from object orientation as well. Users benefit indirectly because easier and more cost effective software development techniques often mean better, more robust, and lower cost software systems. A more direct benefit of object orientation for the user is the ease-of-use that often results from an object-oriented interface. Software objects often model objects from the real world. A well-designed software object will take advantage of any knowledge the user may have about the real-world object by meeting the user's expectations with regard to the characteristics and behavior of the object. For example, a paint program may have a user interface object that models a paint brush.

For a multimedia system, object orientation is very nearly a requirement. Multimedia systems and applications combine multiple media elements in a wide variety of ways. Object-oriented design and componentized software make it possible to design and build a collection of multimedia-processing objects that can interact with one another as well as with the application. As we have seen, many multimedia applications are collaborative, meaning that their primary goal is the sharing of information, including multimedia information, among a number of individuals. We have also seen that many multimedia systems are being designed to support the notion of workflow, meaning that multimedia data must be able to flow from one process to another, and from one software tool to another. More and more, multimedia applications are being designed for distributed environments where software components must be able to communicate and coordinate their behavior across machine boundaries. This allows us to configure the multimedia system according to needs of the application, while the multimedia system components themselves take care of the details of multimedia-data processing. If the underlying object management system supports distributed objects (and distributed resource management) distributed multimedia applications are relatively straightforward extensions of local ones.

Architectural Framework

A framework is a conceptual tool for understanding and describing a complex system. A framework aims to look at a system in terms of the whole problem domain, thereby providing a context for the components of the system. In this section, we will examine three frameworks, each of which deals with a complete multimedia system from a particular point of view. We shall illustrate a multimedia system from an architectural point of view, from a conceptual point of view, and finally from a programming model point of view.

One example of an architectural framework is provided by the Interactive Multimedia Association.[26] The IMA framework is intended to encompass three application domains: stand-alone systems (e.g., games, titles), distributed systems (e.g., teleconferencing), and broadcast systems (e.g., video-on-demand). The IMA model attempts to:

- Separate Control Flow from Data Flow

- Provide Abstraction of File/Data Formats

- Provide Abstraction of Physical Devices

- Separate User Requirements from Application Requirements from System Requirements

By the term data flow, we mean the path the data take through the system from source to destination (sometimes called a sink). A data flow can be viewed from the level of the physical system (i.e., the path the data take from an input device to storage device, or a storage device to a rendering device) or it can be viewed from the level of the logical processes that operate on the data as they move through the physical system. By control flow, we mean the path (or perhaps more correctly, the mechanism) by which a command or instruction is dispatched within a system.

The separation of control flow and data flow can be illustrated by examining a tape player such as a VCR or cassette.

---

26. Corman, Phil, and James L. Green, "Interactive Multimedia Association Architecture Reference Model," *IMA Proceedings*, November 1992.

The data, which in this case are stored on a tape reel, flow from the source reel to the take-up reel, bringing them in contact with a rendering device (the tape heads) along the way. The transport mechanism that turns the spindles that cause the data to flow is under the control of the user through the buttons on the control panel of the tape player. Press play and the spindles move, the data flow, etc. Press stop and the spindles stop moving, and the data stop flowing. Notice that the control commands happen as discrete events, whereas the processing of the data is continuous. This is characteristic of continuous media.

Separation of control flow and data flow for multimedia systems permits the system designer to optimize the data path(s) for efficient and timely movement of large quantities of continuous media, while simultaneously optimizing the control flow path(s) for fast, efficient processing of discrete events.

Another benefit of the separation of control flow and data flow is that it enables the system designer to create abstract components for processing multimedia data that respond to the same control language, but that operate on different kinds of media. A good illustration of this principle is the laserdisc player. A number of consumer laserdisc players can play either videodiscs or compact audio discs, but, in either case, play is still play and stop is still stop. While the example might be somewhat trivial, the situation for digital media is not. Anyone familiar with digital video understands that there is a wide variety of encoding methods currently in use (e.g., MPEG, Motion-JPEG, Indeo, Cinepack, H.261, and so on). The same is true of digital audio. But while it is true that each of these encodings is processed in a slightly different manner, it is also true that each follows a similar processing sequence (which we shall examine more closely later), and that the control language used is largely independent of the encoding method.

It should be clear from the above discussion why the framework includes the concepts of file/data (encoding) format abstraction, and physical device abstraction. The purpose of the encoding abstraction is to emphasize the common attributes of various classes of media, while hiding or isolating (encapsulating) the differences. The purpose of the physical device abstraction is to emphasize the common control commands used to communicate instructions to a media-processing device, while hiding (encapsulating) the specifics of how the processing takes place, or on which vendor's product.

The separation of user requirements from application requirements and from system requirements is essentially an admission that no single conceptual model is sufficient to meet the needs of anyone working with the system at any level. The developer of a system (such as an operating system) thinks in different terms (i.e., that of the systems physical architecture) than the developer of an application. Just as the application developer attempts to create an interface for the user that meets the user's needs and is intuitive and easy to learn, the operating system designer similarly attempts to provide a system interface that is intuitive, easy to learn, and meets the needs of the application developer.

In the IMA model (Figure 7.2), the separation of control flow and data flow also manifests itself in the separation of the model into two main components, the multimedia platform and the multimedia digital data-stream interface.

The multimedia digital data-stream interface is shown in the figure in two ways, as the solid black lines connecting the multimedia platforms, and in an expanded view showing the various classes of media under consideration. These media types include:

- Text   ASCII/Fonts
- Graphics   Computer Generated 2D and/or 3D
- Images   Scanned and Digitized Bitmaps
- Audio   Digitized and/or Compressed Audio, MIDI
- Video   Digitized and/or Compressed Video
- Control   Embedded Control and Attributes in Digitized Streams

As has already been mentioned, each of these media types has a specific structure or format which we have

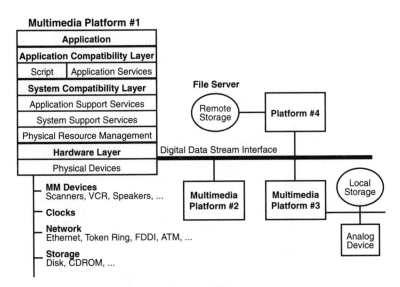

Figure 7.2   IMA Architectural Reference Model

referred to as its encoding. The encoded media, in its pure form is sometimes referred to as a bitstream. These bitstreams are subsequently packaged in various ways for storage/retrieval and transmission purposes. Therefore to complete our discussion of encoding abstractions, we must provide a description of this packaging concept.

To start with a simple example, a bitstream may be packaged for storage as a single stream by placing a header structure at the beginning of the file, followed by the bitstream itself. The header would identify the media type, its encoding method, sample rate, and so forth. A more complicated structure arises when we wish to package several types of media into a single file. In this case, we need a file structure with a sequence of "chucks," each of which contains a header and a bitstream. Since the headers identify the media types, files of this kind can hold a variety of media and are sometimes called compound document files. When a compound document file contains continuous media elements that must be synchronized (as with a digital movie) the ordering of chucks within the file is of extreme importance. The audio and video data which are to be rendered together must be read into the system in a timely manner. It is, therefore, not generally sufficient to place one large chuck for the audio followed by one large chuck for the video into a file.

The solution to this problem is to create compound document files that have a special internal structure called interleaving. Interleaving is the placement of a section of one kind of data with a corresponding section of another kind of data (one section per stream), followed by the next section of each and so on until the end of the file. For the digital movie, we can suppose two streams, one for audio and one for video. An interleaved digital movie file might be packaged as a chuck containing a frame of video followed by a chuck containing a frame's worth of audio samples (i.e., if the video is sampled at 30 frames per second, 1/30th of a second of audio data would be included in each audio chuck).

There is one final data format issue that comes into play when we are dealing with multimedia in a distributed environment related to transmission of data. There is additional information required for data transmission that must be included (e.g., the address of the recipient, etc.), and there are requirements imposed by the underlying network technology with regard to the packet size and transmission latency. Networking software generally handles this level of formatting automatically, but due to the time-critical nature of multimedia data, great care must be taken on the part of the network systems designer to insure that the overhead of packaging and transmission is within tolerable limits.

The IMA multimedia platform is divided into four layers. At the top there is the application itself. Below the application is a layer referred to as the Application Compatibility Layer. This layer is intended to provide the application developer with a set of services that facilitate the creation of multimedia applications. These services may take the form of an application-programming interface (API) or they may take the form of a scripting language interpreter, or perhaps even a high-level application development tool such as an authoring package. Regardless of the form, the goal of the services in the application layer is the same, namely to support the needs of the application developer.

Below the application compatibility layer is the System Compatibility Layer (sometimes referred to as the Orchestration Layer or sometimes as middleware). It is the goal of this layer to provide services for accessing and controlling the devices and processes involved in the manipulation of the data. This layer is also responsible for providing a means of synchronizing the presentation of the media types with respect to one another and with respect to time. Services in this layer generally take the form of a programmatic interface (i.e., a library of functions or a collection of objects). Below this layer is the physical hardware, including any specific multimedia-processing devices.

Conceptual Framework

The main goal of the conceptual framework is to facilitate understanding. "Understanding a thing is to arrive at a metaphor for that thing by substituting something more familiar to us. And the feeling of familiarity is the feeling of understanding."[27] So a conceptual framework is a metaphor for describing a system. Early attempts at selecting a conceptual framework for multimedia systems focused on the system as player metaphor,[28] and at the higher levels of abstraction (i.e., in the application compatibility layer) this metaphor seems appropriate. An application developer that wishes to control a digital movie would like to have a system component that is, in a sense, a virtual VCR. At this level of abstraction, there would exist operations for loading the movie into the machine, and issuing commands for play, pause, stop, etc. The main issues that must be dealt with in extending this metaphor to encompass a complete multimedia system are two: (1) how to generalize the notion of a *player* from the specific media players that exist in the real world (i.e., VCRs, CD Players, Cassette Decks, etc.) to a generalized and configurable virtual multimedia machine made possible by the

---

27. Jaynes, Julian, *The Origin of Consciousness in the Breakdown of the Bicameral Mind*, Houghton Mifflin, 1976, p. 52.
28. For example, the Super-VCR model used in the original DVI system.

programmability of the digital system; and (2) how to decompose the system into more primitive components that enable these virtual multimedia machines to be constructed.

In designing the second generation DVI system, a conceptual framework was developed called the digital video production studio.[29] The basic idea was to provide components modeled after those found in a typical production studio. This included mixers, tape decks, monitoring systems, effects processors, and various other items that one could connect together to record, modify, and play audio or video tracks. In the DVI model, a digital production studio is composed of an analogous collection of software/hardware subsystems that operate on digital streams of audio and video data. A collection of streams that are intended to play together forms a "group." Analog inputs and outputs can be thought of as "channels." Streams of audio and video data are routed from input channels to output channels by making "connections." Streams can be "mixed" together (e.g., assigned to the same output channel) or they can be routed through an "effects processor" to alter the data in some way.

Using the conceptual model described above, objects were identified and abstracted to form the basis for the AVK interface. The interface consists of a collection of objects with their associated behavior and attributes.

The analog interface subsystem (patchbay) contains two objects, the AVK *Session*, and the AVK *Device*. The calls associated with these objects start up AVK, define the communication mechanism between AVK and the application, allow query-of-device capabilities, and open and close the DVI device within an AVK application session.

The Stream Manager (tape deck) is implemented as a collection of objects that control digital data streams. These objects were treated by AVSS as a tightly coupled collection (i.e., a file). AVK treats these objects as independent of the file (or files) storing the data. A *Group* is the unit of control

---

29. Green, James L., The Evolution of DVI System Software, CACM, Vol. 35, No. 1., January 1992.

synchronization and communication (i.e., the tape transport controls). *Group* calls include functions such as play, pause, and record. A *Group Buffer* represents the tape. A group buffer can contain multiple *Streams*, all of which play at the same rate. A *Stream* is a single track of audio or video data. Several streams may be stored in the same file.

The function of the mixer in AVK is provided by an object called a *Connector*. A connector can be thought of as a pipe which accepts data flow in, optionally transforms those data, and pumps data out. The connector is a higher level abstraction of a copy operation. Connectors allow rectangular regions, called "boxes," to be defined for the source and destination bitmaps. The size of the boxes can be modified in real time to allow resizing and relocating of images to support windowing.

The display subsystem is embodied in an object called a *View*. A view is a displayable image and a collection of visual regions (boxes), which are mapped into windows by the application. A view is most often the destination of a connector where the source is another displayable object such as an image or a video stream. An application can switch among multiple views.

The sampler in our current design consists only of the *Image* and *Image Buffer* objects. An image is a portion of VRAM which may be used to store stills. An image buffer is a compressed image. Images can be "compressed" into an image buffer, and image buffers can be "decompressed" into images. Image effects can be performed in place on an image, or as part of a copy operation using a connector.

Programming Model

At the next level of detail is the programming model. An excellent example of a programming model for multimedia is described by Gibbs.[30] He presents a programming model based on multimedia-processing objects that can be con-

---

30. Gibbs, Simon, "Composite Multimedia and Active Objects," *OOPSLA '91 Proceedings*, ACM, 1991, pp. 97–112.

nected together to process composite multimedia, which he describes as "a multimedia object containing a collection of component multimedia objects and a specification for their temporal and configurational relationships."[31]Assuming that a variety of such composite multimedia objects are possible, the software machine that must be constructed to render them must be configurable, or must allow for the construction of virtual multimedia machines that can be dynamically constructed out of a set of media-processing objects. In the Gibbs model, media flow between these media-processing objects from sources to sinks with optional filters between for performing transforms on the media prior to their being rendered at the sink.

Another example of a programming model that is also based on the notion of connecting media-processing objects together to form virtual multimedia machines is given by the IMA's "Multimedia System Services" specification.[32] The programming model is very similar to the Gibbs model. It defines a collection of media-processing objects called *virtual devices* that can be connected via ports into graphs, which represent the multimedia machine required to render the various media elements that comprise what Gibbs would call a compound multimedia object (Figure 7.3).

Models like this are very powerful in conceptualizing a system design and defining the interfaces and implementation of flexible software services in support of multimedia processing. They enable the software engineer to define the problem domain and solution space (the framework); to provide a description of the system in terms of a relevant metaphor as an aid to understanding (the conceptual model); and to provide a structure as a guide to detailed systems design and implementation (the programming model). Beyond this point you get into the specifics of the interfaces and the implementation of the system services.

---

31. Gibbs, p. 104.
32. "Multimedia System Services, Version 1.0."

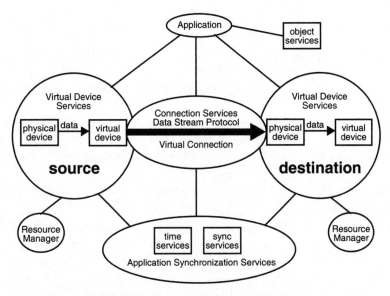

Figure 7.3   IMA System Services Connection Model

Applicable Standards

Absence of Dominant Standards in Multimedia Software

Lack of software standardization has often been mentioned as a stumbling block in the deployment of multimedia technologies, but the debate over the specifics of exactly what needs to be standardized and who should set the standards often confuses more than clarifies these issues. I mentioned in a previous section that object-oriented techniques are most directly beneficial to the software developer and only indirectly beneficial to the consumer. If we take the benefit to the consumer to be our primary criterion, then it is interoperability standards that are the most important. Interoperability, in this sense, is the ability to communicate and share data with other consumers, even if they are using different platforms. A good example of this is the modern telephone. You can buy your phone from anyone, but it will plug into the system and connect you with anyone else on the system, regardless of who they bought their phone from.

Standards in other areas, such as those associated with programming interfaces for creating multimedia applications, are more like object-oriented techniques. They would most directly benefit software developers and only indirectly benefit consumers. For the former group, interface standards might make it easier to develop cross-platform applications, which would indirectly benefit the latter group by bringing more cross-platform applications to market. There is another side to this debate, of course, which argues that platforms compete by differentiating themselves, and that some application developers will differentiate their own products by taking advantage of these platform specific features. This creates a dynamic where innovation and market forces can serve to push the state of technology forward faster than might otherwise occur. I do not wish to enter into this debate here, but would rather make the point that interoperability standards should be of more interest to consumers than software interface standards.

Data Standards for Audio and Video

Before audio and video data can be stored and manipulated by digital computer systems, they must first be encoded. In other words, a code must be developed that provides the programmer with a representation of the original information in a form that can be easily manipulated by the computer hardware and software. For traditional data types such as letters and numbers, these codes are fairly well standardized. Definitive digital representations of newer data types such as images, audio and video, have yet to be determined, and a number of competing representations exist.

There are several reasons for the lack of standard formats. One is our lack of experience with digital representations of audio and video. Another stems from the fact that the people inventing the codes generally attempt to find optimal representations tailored to the needs of their specific market. Immediately, there are two different perspectives, one involving the integration of audio and video, as data, into the computing environment, and the other involving use of the computer as a tool in the production of audio and video to be used in traditional media environments such as television. The needs of these two classes of user are (at the moment) quite different, and the result is that different representations have emerged to meet those needs.

The concept of an optimal format revolves around three basic issues: efficiency of storage, speed of encoding and decoding (which are often different), and the level of quality desired. Different users of audio and video information make different trade-offs among these three attributes. For example, in the world of video-conferencing, it is more critical that the encoding and decoding happen in real time than it is to have the best quality audio or video. For video production professionals, it is more important to produce the best quality possible, even if it takes longer to encode, whereas publishers of multimedia titles may wish to encode the data so that they can be efficiently stored and retrieved off a CD-ROM even if it means a slightly lower quality.

To complicate things still further, these data types are often augmented with additional information (beyond the audio and video data) in the form of *file headers*, and *data set descriptors*. This additional information can be used to identify data types, or specify parameters that were set when the data were created, and/or that is required for them to be properly rendered. In many cases, more than one type of data is stored in a single file. Files that can hold more than one type of data are often referred to as *container files*. As with data types, the formats of container files can vary from one computer or operating system to another, causing problems whenever the user wants to move the data between different systems.

A final reason for the lack of standard representations of digital media is that computer and software companies are not used to thinking in terms of standards in anything like a universal sense. In the computer industry the idea of a standard is something invented by a particular company for use on a particular platform (or selection of platforms), which the company encourages people to use (sometimes for a fee). This tradition stems from the competitive advantage gained by having the user's data encoded in a form that can be processed efficiently by one company's system, but which must be converted or re-created before it can be used on another. This effectively creates a form of inertia that prevents users from switching to a competitive product. The file import/export options available on many application programs are in fact an attempt to overcome this inertia. This trend has been affected in recent years by the emergence of heterogeneous network environments where different computers running different software systems are being linked together and must be made to share data and to communicate with one another. Nevertheless, there will continue to be different encoding schemes for different uses and from different vendors for quite a while (maybe forever).

Industry Outlook

I am going to take some liberties with the concluding section of this chapter for the following reason. If I were to try to make predictions, I would strive to make accurate ones. In other words, I would try to predict what I think is most likely, given the current state of affairs. If I were to do this, I'm afraid I would be somewhat pessimistic about our future. My chief concern here is that logistical, economic, and government regulatory issues will cause us to fall short of the potential offered by the digital convergence. So instead, I am going to attempt to draw attention to the potential. Therefore what follows is the industry outlook as I would like to see it, rather than as I actually see it unfolding before me.

At the beginning of this chapter, I introduced the notion of the universal machine, which I define as a general purpose, programmable, communications and computing device. In conclusion, I add to that the notion of the *universal man* as embodied in such men as Leonardo da Vinci and Isaac Newton. These men saw few boundaries between their various disciplines and were able to wander (browse?) effortlessly through all the different fields of knowledge available to them. These men and others like them did not simply generate new information, but instead performed a synthesis that resulted in useful knowledge. They were able to do this primarily because of the advent of a new means of communication called the printing press, which enabled the creation of libraries and the critical exchange of ideas on a scope that was impossible to imagine in the generations that preceded them.[33] But these Renaissance men did not spring into our history from thin air. They were, in fact, the beneficiaries of that preceding generation.

On the cusp between the Middle Ages and the Renaissance there was a movement known as Humanism. This movement focused attention on issues related to human beings, and was founded on a fundamental belief in human potential, particularly the potential of the human mind. One of the outcomes of this movement was the writing and translation of texts in vernacular (as opposed to scholarly) languages. This was essentially a democratization of information. It was this very access to information that enabled the synthesis and advancement of knowledge that occurred during the Renaissance. We have among us today those who espouse theories and conduct research into human–computer interaction and who evangelize the notion of user-centered design—which is, once again, signaling a kind of renewed humanism. It is essential that we understand the magnitude of the potential that lies before us. If we combine our focus on human user interface design with a fundamental belief in human potential, if we see our task as one of removing barriers to knowledge and to communication, if we strive for the democratization of information, we will be contributing to a movement which has the capability of sparking a twenty-first-century renaissance. We, like the humanists of the thirteenth century, stand on the cusp of what could be the next great age of mankind. Let's not sell it out for the sake of mere entertainment.

---

33. Green, James L., *A Lesson in History for the Digital Revolution*, NAB Multimedia World.

8

ATM Deployment: Architectures and Applications

HARRY L. BOSCO

Say "networked multimedia" and most people will respond "ATM." Rarely have so many uniformly agreed on the probability of the application on one network technology: Asynchronous Transfer Mode. As the leading data-networking transport mechanism, it is appropriate to look at how and where ATM is likely to be deployed.

The costs and complexity of ATM are still relatively high. As a result, it is likely to be rolled out in a staged fashion where its unique high volume characteristics will yield immediate payback within existing telecommunications networks. This chapter explores the likely introduction sequence and the surrounding environmental drivers.

Overlapping phases of deployment are described, starting with the deployment of basic broadband technologies over which ATM will later be introduced, followed by the integration of ATM with current switched networks. The orderly and sequential introductions of such network capabilities may frustrate those who have hoped for networked multimedia capabilities sooner rather than later, but, once again, the complexity of the pre-existing infrastructure makes fast transitions impossible.

The regulatory situation could, however, serve to speed up adoption processes if telecommunications companies are finally allowed to compete with cable companies and provide new multimedia services. Deregulation is likely to provide the business incentive, in the form of potential return on investment, necessary to accelerate ATM implementation.

Deregulation notwithstanding, ATM technology is not cheap. ATM-capable CPE (Customer Premise Equipment) is, for now, more expensive than can be justified in today's market. Alternative transport mechanisms such as MPEG 2 (in common with consumer electronics, broadcast, and cable television plans) are being viewed as the most likely transport model into the home itself until an ATM-based infrastructure becomes cheaper. Carriage of MPEG 2 over backbones will almost certainly utilize ATM, however.

While there is general agreement on the value and utility of ATM, this chapter describes an implementation scenario that may be less popular than most would like, but probably more realistic than many have been predicting. Once again, the complexities and history of the industry govern the evolution more than is often realized.

Introduction

Today's business and consumer communications end users require the movement and management of an increasing quantity and variety of information. This information explosion will shape the evolution of communications networks and services in ways which we cannot fully comprehend. Network equipment suppliers and service providers are challenged to develop and deploy networks whose cost would effectively meet the current needs of their customers, while simultaneously evolving to address their future needs. Network service providers worldwide are making considerable progress in the deployment of narrowband ISDN and

Advanced Intelligent Networking. Yet despite this progress, these technologies may only partially address the needs of leading-edge customers who demand increasingly sophisticated real-time interactive multimedia services.

Asynchronous Transfer Mode (ATM) technology and networks have a clear potential to alter today's communications infrastructure dramatically, in less than a decade. By deploying ATM equipment, service providers will be positioned to deliver orders of magnitude more bandwidth to customers flexibly, consolidate central office equipment functions, and vastly simplify operations, administration, and maintenance of the network. This is, however, only part of the potential payoff. An even greater potential for revenue enhancement can be had by network operators who deploy network architectures that lend themselves to rapid and open service creation environments. New imaginative applications and services must drive the growth of network usage, bandwidth demands, and revenues. By providing a network architecture that allows service providers to rapidly respond to new service concepts, new services and innovations can be quickly brought to market.

ATM Deployment Drivers

Many forces will shape the evolution of today's shared service provider networks. These forces include the evolving needs of the end-user, the challenges and opportunities facing shared service providers, and the evolving technology and standards environment. Key forces within these areas include:

- End-User Needs

  End-users are beginning to demand greater control and flexibility over their communications resources. In particular, end-users require flexibility in matching a grade/quality of service to application needs, ability to quickly allocate and reallocate bandwidth to particular tasks and capabilities to integrate voice/video/data for multimedia solutions.

- Service Provider Challenges

  Service providers today face many significant challenges and opportunities. Given changing regulation and intensifying competition in many parts of the world, service providers must protect their existing offerings by adding value and by reducing costs. Service providers face both the challenge of new competitors as well as the possibility of intelligence migrating out of the network to the customer's premises. In addition to protecting their current offerings and revenue sources, service providers must be able to grow revenues via the rapid introduction of new services and capabilities. This will require the introduction of new technology that will allow for quick service-creation as well as technology that can support the coming evolution to multimedia services. Protecting and growing the revenue base must also be accompanied by the overall reduction in the cost and complexity of network operations.

- Evolving Environment

  Several key industry trends will also have a significant impact on the overall evolution of public shared networks. Synchronous Optical Network (SONET) and Synchronous Digital Hierarchy (SDH) will be deployed as part of the overall modernization of public networks. The deployment of SONET/SDH, in and of itself, will have several benefits such as improving operations capabilities and increasing the overall reliability of the network. In addition, the deployment of SONET/SDH will be an enabler for the introduction of high-bandwidth ATM services that will ride on top of the SONET/SDH transmission infrastructure. A second key trend is the emergence of global standards for ATM/BISDN and the convergence of the computer/CPE industry on ATM as a next-generation communications technology. Global standards will insure the ability of end users to get seamless services on a multinational basis. In addition, the adoption of ATM by the computer/CPE community will reinforce the need for ATM within the public network to allow for efficient communications between private network equipment and the public network. A third key industry trend is the move to multimedia services and the concept of the information superhighway (also called the Infobahn).

ATM Deployment Strategies and Evolution

While it is not possible to predict the exact evolution of today's shared public network to ATM/BISDN, it is likely that it may evolve in four overlapping phases. In the first phase (essentially today) the shared public network consists primarily of circuit-switching equipment supporting telephony and private line services. Packet-switching equipment, if present, is supported as an independent or overlay network. Within this time frame, service providers are deploying broadband technologies such as SONET/SHD into the network. In addition, service providers may upgrade their packet data networks to support fast packet services such as Frame Relay and SMDS. These activities will pave the way for later introduction of ATM and BISDN. Therefore, we refer to this phase as the broadband introduction phase.

The next step in the deployment of ATM/BISDN will be the introduction of overlay ATM-based networks. In this phase, new ATM equipment will be deployed to create overlay networks, adding new broadband capabilities to augment the existing circuit-switching network. These overlay networks will grow and evolve along with the growth and evolution and broadband services and applications. We refer to this phase as the ATM overlay phase. As the overlay ATM/BISDN network grows, the service provider will find it advantageous to integrate the operations of the ATM/BIDN network and the circuit network to gain greater simplicity in operations and to reduce the overall cost of service delivery. In addition, given the need to communicate between the two types of networks, service interworking will be required. This phase represents the integration and consolidation phase of ATM/BISDN evolution. In the final (?) phase of evolution the integration between the ATM/BISDN network/capabilities and that of the circuit network have merged into a cohesive and consistent network architecture providing a broad spectrum of voice, data, video, and multimedia services. We refer to this phase as the unified network phase.

Broadband Introduction Phase

Many network operators have been making changes to improve the capabilities and infrastructure of their current communications networks. These changes are targeted at permitting the introduction of new fast packet services such as Frame Relay and SMDS/CBDS as well as at improving the overall quality and reliability of the network. While these improvements are not directly targeted at providing ATM/BISDN services, they can pave the way for the introduction of ATM/BISDN in the future.

One overriding aspect of network evolution is the modernization of the transport network. Much of the new transport infrastructure being put in place today is based on SONET/SDH technology. While the deployment of this technology is motivated by a variety of issues (improved maintenance, self-healing, increased bandwidth . . .), it also has the associated advantage of being the transport infrastructure of choice of ATM/BISDN implementation. Thus the introduction of SONET/SDH as part of a network modernization/upgrade program will serve as an enabler for the introduction of ATM/BISDN.

In addition to the introduction of SONET/SDH, many service providers are currently upgrading/evolving their packet networks from X.25 and related systems toward fast packet technology. These services, such as Frame Relay and SMDS, are closely related to ATM/BISDN concepts and will also provide a proving ground for and a stepping stone toward ATM/BISDN. Deployment of these new services and of fast packet technology will provide needed experience for the next step in the introduction to ATM. In addition, growth of services such as Frame Relay and SMDS/CBDS will lead many service providers to consider ATM as a consolidating backbone technology for all their packet services.

ATM Overlay Phase

The ATM overlay phase begins as the service provider deploys equipment to provide direct ATM services. Given the significant embedded base of primarily (STM/PDH) circuit-based equipment, the introduction of ATM will most likely be accomplished via creating an overlay network. This overlay network will form the basis of initial Permanent Virtual Circuit (PVC) offerings to end customers. In this stage, the ATM equipment forms its own network or subnetwork and is operated and managed separately from the main network. The ATM network may share resources with the main network such as transport transmission facilities, but the operations and control remain independent from today's voice/telephony capabilities. This overlay approach is most efficient for the rapid introduction of basic ATM services without requiring significant changes to the much larger voice network. This allows service providers to deploy ATM services rapidly and simply. The development of overlay networks also provides a training ground for service providers and network operators to gain experience with ATM before integrating it with the main voice network.

Throughout this stage, there will be an important link between the overlay ATM network and the main voice-oriented network. This major link is the use of SONET/SDH facilities to transport both ATM and circuit-oriented traffic. By sharing facilities, the service provider gains the benefit of more efficient utilization of the transport infrastructure and avoids the problem of needing to operate two independent sets of transmission facilities.

ATM Integration and Consolidation Phase

As the volume of ATM-based services grows, the need to begin to integrate the overlay ATM network and the circuit-based network will be fueled by two main issues. The first of these is the desire of the end user to get integrated services incorporating the endpoints and features from both the ATM and circuit networks. A typical example of this is the desire to link BISDN and NISDN endpoints together into a single audio or video conference where each endpoint device can participate, limited by its capabilities (i.e., narrowband endpoints will not be able to receive 45 Mbps video!). The second issue which will drive the integration is the need to have coordinated operations and management of both networks. Unless some level of operations integration is achieved, maintaining and provisioning both networks separately (especially when integrated services are required) will become complex and costly.

These issues will drive the integration of the ATM and circuit networks closer together. As ATM becomes a larger part of the overall network, and as it grows more entwined with other services, network economics will dictate closer and closer integration of the overall communications network. As ATM assumes a larger role, the ATM infrastructure will begin to take on additional capabilities such as advanced voice services and as a unifying network for access and transport.

Unified Network Phase

The logical end result of this evolution may well be the emergence of a single unified ATM network supporting voice, video, and data services. While it is unlikely that this phase of evolution will occur in the next several years, it does provide a target for network evolution planning. Given the inherent capabilities of ATM, this unified network concept could provide a broad spectrum of services across a wide range of bandwidths.

Broadband Application Evolution

As described above, the public network will evolve in a series of stages; so, too, will the applications and services supported evolve. Initially, ATM capabilities may be used to support specialized services and applications such as providing a backbone for Frame Relay and SMDS. Gradually, specific ATM services such as ATM private line will emerge. Finally, as new applications are developed and network capabilities mature, multifaceted switched ATM services will emerge. The starting point for this evolution is the combination of today's communications networks, available telecommunications standards, and near term demand for business data services and residential video services.

Business-Market Evolution

In the business market (assumed to include work-at-home situations), ATM services will initially augment the currently available set of voice-networking and data-networking services. Today's business networks carry a wide range of applications, which can be sorted into various classes. These classes include voice service, low-speed terminal data communication, higher-speed local-area network (LAN) interconnect and LAN backbone transmissions, imaging data services, bulk data transfers, and video teleconferencing sessions. These applications are carried over a correspondingly wide range of communication networks. Typically, each application structures its own subnetwork around a communication-network service that is best matched to its needs.

Current market-driven applications—LAN interconnect, collaboration using high-resolution imaging data, and video teleconferencing—have hastened the introduction of several new high-speed services, based on relatively mature circuit-oriented and packet-oriented technologies. As new ATM-based customer-premises equipment and networks capabilities become available, demand for semipermanent (provisioned) ATM public-network services will emerge.

These network service offerings will enable end-users to design private and virtual private networks at broadband rates. New services offerings will coexist with today's non-switched and switched network services. The initial use of ATM services in this environment will be to extend further the available set of networking options so that high-speed packet-like applications, including LAN interconnect, LAN backbones, and high-resolution image transfers, can take advantage of ATM's efficiency, high transmission speed, and ability to handle "bursty" data traffic. In some cases, existing application networks will move to ATM, possibly using protocol adaptations between existing applications that were not practical earlier (because of cost or performance limitations), but will become possible due to the availability of ATM services.

As applications move to ATM, ATM-based networks could be used to consolidate multiple, separate data networks. These applications will take advantage of the fact that many carriers and equipment manufacturers will transport communications protocols, such as Frame-Relay and SMDS services, over ATM backbone networks. By taking advantage of protocol adaptations, current applications (using Frame-Relay and SMDS services) as well as new applications (running over ATM services) can be combined onto a single data-networking environment based on ATM. Eventually, as multimedia applications become more prevalent in data networks, many will evolve to employ the ATM protocol directly. Over time, this will reduce the need for protocol adaptations.

As broadband ATM becomes more cost effective and ubiquitous, end-users will take advantage of ATM's ability to carry many different types of traffic, and will extend the trend toward network consolidation to areas beyond data networking. For example, the ATM backbone, which will initially be deployed to support data-networking traffic, can readily be used to transport high-speed circuit payloads, such as video traffic and voice trunks, at 1.544 Mbits/s (DS-1 rate) or higher. This consolidation trend will lead to the use of ATM-based, integrated, wide-area-network access. Integrated access involves converting all of a customer's communication services to ATM at the customer's premises, and then carrying them into the wide-area network by means of a single, high-speed ATM access line. Each service may still be treated as a part of its own sub-network, but the common ATM access line allows transport efficiency and bandwidth sharing among the services and subnetworks.

As this continues, the desire for on-demand (real-time switched) networking capabilities will emerge. The addition of intelligent switching capabilities to the network will enable the introduction of many new and useful features. First, the required bandwidth can be allocated and deallocated on demand and in real time. Second, connections beyond a company's private or virtual private network can be established when and where they are needed around the globe. More significantly, features now available for voice services through the intelligent network can be applied to broadband ATM services, such as "700," "800," and "900" switched services, call screening, and call forwarding—to name a few. All the while, the more "bursty" traffic data and image will be statistically multiplexed to maximize bandwidth, thereby increasing transport-network facility utilization.

Over time as multimedia applications become more prevalent and as other applications increasingly demand flexible service capabilities, the multiple services provided over integrated access lines will evolve toward fully integrated network solutions. In an integrated network environment—a wide range of services with various levels of performance and intelligence will still be available. However, these services may be offered in selectable grades within a single, integrated networking environment—the resulting integrated communications environment. That environment will provide the full service and bandwidth flexibility needed to meet, most efficiently, the diverse and rapidly evolving applications that will likely characterize the progressive business environment.

Residential Market Evolution

Advanced residential information services have achieved only limited market penetration to date. On the other hand, with the trend toward more leisure time spent at home, residential video-entertainment services have been very successful. For reasons that follow, ATM technology will play a key role in the evolution of advanced video-entertainment services. This may provide the stimulus for widespread demand for residential, broadband data and multimedia services, in addition to video-entertainment services.

While today's video-entertainment delivery systems are based on analog technology, there is considerable consumer demand for higher quality and more reliable CATV services. Consequently, the industry is quickly moving toward systems now in development that will use digital channels and rapidly advancing, digital video-compressing technology. These systems will provide brand-new capabilities such as video-on-demand, high definition television (HDTV), and 500-channel CATV.

In several of these capabilities, multiple, compressed, digital video channels are carried over a single digital bit stream. Digital video-compression technology permits encoding video signals at variable bit rates. In many applications, a single bit stream may carry multiple digital channels running at different bit rates (various aspects of the program material can affect the desired video bit rate, such as the amount of motion, and whether the original source material was from film or videotape).

Within a telecommunications network, ATM technology is ideally suited to serve as the multiplexing technique for carrying multiple, compressed, digital video signals over a single digital bit stream to the home. ATM technology is highly effective in carrying multiple channels of arbitrary bandwidth over a single digital bit stream. In addition, ATM is well suited to the additional flexibility demanded by some digital video-entertainment applications—in many cases, a single video-entertainment application. In many cases, a single video program can be associated with multiple audio signal (multiple languages) or data overlays (closed captioning or ordering information). All these signals and overlays can be carried efficiently as ATM virtual circuits in a single, multiplexed bit stream.

The value of ATM in carrying compressed digital video is reinforced by recent video-compression standards work from the Motion Picture Experts Group (MPEG). The MPEG-2 standard has been designed to enable MPEG-2 compressed video information to be transported via ATM calls over an ATM network.

Initial residential use of digital compressed video will be for CATV systems with an extended number of channels. Early services will probably include an expanded channel selection, a wider range of pay-per-view events, and an enhanced pay-per-view system, which might provide the most recent movie releases at 15-minute intervals. These services will be followed by video-on-demand service, which will use the extended number of channels to provide interactive and fully controllable viewing of movies supplied by video servers.

The network and customer premises equipment used to provide these video-entertainment services includes key underlying capabilities, which facilitate the deployment of more advanced residential services. The underlying capabilities of a CATV channel-expansion system with video-on-demand include:

- Broadband channel to the residence,

- The ability to select and process broadband digital channels, and

- The ability to transmit information in the "upstream" direction from the residence into the network.

With the flexibility of ATM, these basic capabilities will help to provide a full range of audio, video, data, and multimedia applications to the home. Applications will include entertainment, education, and telecommuting services. Thus, the initial deployment of ATM video-entertainment services will lead to the widespread deployment of high-bandwidth equipment for the home. This equipment will have the flexibility to offer the sophisticated capabilities expected by today's media-conscious residential consumers.

Fiber-based and coaxial cable based distribution systems, which are being installed today, can deliver 50 to 80 analog channels to communities ranging in size from a few hundred homes to a few thousand homes. With the technology described earlier, these types of installations could provide a dynamically "sharable" bandwidth of 1.5 Gbits/s to 2.4 Gbits/s (or more) to each community. These broadband distribution systems, when combined with the bandwidth flexibility provided by ATM networks, create an astounding array of opportunities for service and application evolution.

Conclusions

The development of broadband ATM services will be driven by end-user application demands. In the business environment, explosive growth in data networking will drive the deployment of ATM campus-backbone networks. Deployment will be accelerated by client and server distributed network computing, and applications utilizing progressively more graphic and video-oriented content. Moreover, the requirement to extend local campus network capabilities to the entire wide-area corporate enterprise will spur demand for initial, carrier-based, ATM-provisioned services. Emergence of cost effective client workstations will encourage the development of many new multimedia applications. And finally, the need to extend the information "superhighway" to corporations' external clients (both its customers and its suppliers) and to its telecommuting employees will usher in the evolution to intelligently switched, broadband ATM services on premises and across public networks.

In the residential market, consumer preferences, new regulatory rules, and technology advances will encourage the development of an exciting array of advanced video-entertainment services to the home. Some of the most advanced video-distribution services have already been field tested and are in early market deployment stage. Consumer demand for interactive video applications will encourage the introduction of early multimedia service offerings. Some of these include interactive TV, interactive tele-education, and perhaps even event simulcasting with tele-education, as well as even event simulcasting with tele-wagering.

Initial success in broadband market development in the business and residential sectors coupled with the technology trends discussed earlier may reinforce each other in much the same way that the cellular telephone market has developed. This market "churn" will further encourage competition between traditional LECs, access providers, and CATV companies, thereby stimulating the rapid introduction of broadband services and capabilities based on ATM technology. Then, as market penetration rises and prices fall, the dream of advanced multimedia telephony services for the mass market can become reality.

At the core of this change is ATM, a technology that can dynamically support a broad range of applications. ATM will provide the communications infrastructure to support the convergence of telecommunications, computing, and video (television). It is difficult to predict when ATM technology will deliver on all its promises, but compelling evidence suggests that ATM will be the technology of choice for the future of the information superhighway.

9

Telecommunications

PHILIPPE CLARKE

Telecommunications shares with broadcast television a long history of government regulation and intervention. These two industries also share the highest-per-household penetration and usage of all electronic markets. Is there a connection between usage and regulation? Yes. It's called public service. The question the government asks is "what is in the public interest?" The answer for telephone services was a tightly regulated monopoly—at least for a while.

Years of monopoly prevented duplication and assured common standards, all of which benefited U.S. consumers. That history, however, created a homogeneous infrastructure that has limitations that retard the addition of new multimedia services. This chapter, however, reviews the sheer complexity of these services within an historical context that helps explain why these networks are not likely to be abandoned too quickly.

Unlike the cable television industry, telecommunications networks have had to operate as a "common carrier," which has been both a blessing and curse. The blessing is that the demand for new services has grown because of the widespread use of facsimile machines and data modems over circuits that were originally designed principally for voice. The curse is that telecommunications companies have not so far been allowed to own any of the information flowing through their system.

New interactive multimedia services should, in theory, be an important driver in the growth of data services since the bandwidth requirements for new media data are huge. But mapping that demand to specific markets and network infrastructures takes time. As with all technologies, the business model for the uptake of multimedia technology in business applications is quite different from anticipated consumer services. The usual consumer demands for higher quality and lower cost versus the business demands for extremely high service reliability promotes the emergence of separate networks for business and consumers, each with their own capabilities.

Both telephone companies and cable television companies run wires into homes. Both are subject to government regulation. And they are potentially direct competitors. The U.S. government usually takes the position that competition is good, and when it exists the need for regulation diminishes or disappears. Clearly the government holds the key to the rate of change in these two markets through the adoption of new legislation that could remove the current restraints on these two giants, unleashing a flurry of activity. Both groups are champing at the bit for the opportunity to protect and enlarge their markets. The question is, will that be in the best interest of consumers?

Introduction

"Keep in touch!" For hundreds of years, the most common way people kept in touch was by sending a letter through the world's postal services. In 1844, businesses and consumers began to use Samuel Findley Breeze Morse's electric telegraph to send telegram messages instantly over long-distance wire. Then in 1876, Alexander Graham Bell introduced the telephone, which laid the foundation for an industry which serves nearly every home, business, and government in the world. Postal and telephone services continue to be the most common method used by businesses to communicate with customers, suppliers, vendors, and employees. This balance will change, however, as more consumers and businesses become aware of the communication options available today.

People can choose to communicate by facsimile, electronic mail, interactive bulletin boards, X.400, EDI, video phones, teleconferences, video conferences, voice mail, and automated voice response systems. New technologies have improved the number of communications channels available for people to send and receive their messages. Businesses can communicate through complex networks which may include one or several of the following technologies: satellite, radio, cellular, ISDN, ATM, Frame Relay, and packet switching.

Whether you work for a local or international business, a city, state, or federal government agency, or as an entrepreneur in your home, the communications options you select will affect your ability to communicate effectively and will influence the overall success of your endeavors. Therefore, a solid understanding of the telecommunications industry, including the products and services provided by common carriers, local regional carriers, suppliers, and value-added service providers, will give a competitive advantage to people trying to start a business, market a product or service, improve information flow, or reduce overall communication expenses.

Industry History

On Friday, January 8, 1982, AT&T agreed to divest twenty-two local Bell operating companies worth approximately $80 billion. This agreement ended AT&T's tight control over competition in the telecommunications industry. As a result, the current marketplace for telecommunications products and services changed from a virtual monopoly to a consumer-driven, highly competitive environment. To better understand our current environment, a short history lesson is necessary.

The First Phone Companies

On March 10, 1876, Alexander Graham Bell was successful in saying, "Mr. Watson, come here; I want you" to his assistant over a telephone transmitter and wire which he invented. This event led to the development of the telephone industry and to AT&T becoming the largest monopoly ever allowed to exist in the United States. AT&T, until recently, owned, operated, and controlled nearly every part of the U.S. phone network including local loops, long-distance services, switching facilities, and transmission gear. But this was not always the case. AT&T did face some stiff competition in the beginning.

Competition came soon after Bell's discovery of the telephone. In November, 1877, Western Union created a subsidiary called the American Speaking Telephone Company. Western Union, with much more financial strength and technical resources than the Bell enterprise at this time, acquired several patents and even hired Thomas Edison to develop telephone equipment. Western Union's equipment and services were far superior, and the rates being charged were much lower than Bell's. The only way the Bell enterprise could compete was to bring a patent infringement suit against Western Union. The suit was settled in November, 1879, with Western Union agreeing to completely withdraw from the telephone business.

At the turn of the twentieth century there were more than 500 independent telephone companies established in the United States. However, these companies did not cooperate with each other or with the Bell interest. They were small telephone network providers which catered to small niche markets and, therefore, they did not care about providing access to other independent network carriers. AT&T understood from the beginning the importance of interactivity and connectivity among all sites in a nationwide telephone network and the complexity involved with expanding this network. AT&T felt justified in having direct ownership of every part of the network including the phone equipment, local loops, switching centers, transmission gear, and long-distance lines. Therefore the corporate direction of AT&T very early in the game was to become a vertically integrated company with the strength to force smaller independent companies to follow their lead and become licensees.

In order to maintain influence with the local licensees, AT&T, known as the American Bell Telephone Company until 1899, offered permanent licenses in exchange for stock in the local company. Gradually these small licensees consolidated into larger territories. This further diminished the influence and control of any individual licensee and made it easier for AT&T to increase its equity holdings. In 1900, AT&T acquired Western Electric, an electrical equipment manufacturer, because it could no longer depend on other manufacturing companies to meet the expanding demand for high-quality telephone equipment. The last component of AT&T's predivestiture structure was the incorporation of the Bell Telephone Laboratories (Bell Labs). Bell Labs was formed to coordinate research and development, planning, engineering, and design.

AT&T was fortunate that the U.S. government felt there would be little advantage in encouraging competition among different phone companies in the same geographical area. Furthermore, it was felt that the duplication of services and facilities would be wasteful and inefficient. AT&T capitalized on this by rejecting interconnection with any independent telephone company that competed against Bell licensees and affiliated independents. Customers on independent telephone networks were often unable to communicate with other people in the same town because of this interconnection issue. The great majority of independents who refused to join the Bell system because they favored competition did not survive because people on their network could not converse with anyone outside the local network.

Divestiture—1982 Landmark Agreement

The 1982 AT&T divestiture agreement entirely changed the structure of the telecommunications industry by forcing the breakup of a monopoly and promoting competition. Prior to divestiture, AT&T owned, operated, and controlled nearly every aspect of the U.S. telephone network. In 1974, the Justice Department filed an antitrust suit accusing AT&T of conspiring to prevent, restrict, and eliminate competition from other private common carriers, manufacturers, and suppliers of telecommunications equipment. Seven years later, an agreement was reached that stipulated that the U.S. Department of Justice would drop the antitrust suit, if AT&T agreed to give up its 22 local Bell operating companies.

The antitrust suit was a deliberate attempt by the U.S. government to stimulate competition in long-distance and terminal equipment. The divestiture now allowed any competitor to enter the field and compete for long-distance and equipment market share. Because of the divestiture, AT&T was also free to enter almost any competitive market except for information services and electronic publishing, which was prohibited until August 25, 1989.

Resulting Effects from AT&T's Divestiture

If the goal of the U.S. government, in causing AT&T's divestiture, was to stimulate growth and promote advanced technologies in the telephone industry, then I believe they succeeded. Consumers and businesses have gained in many tangible ways such as improved customer service and new pricing schemes. Nationwide television commercials from AT&T, MCI, and Sprint constantly advertise various new pricing schemes such as "Friends and Family," "Reach Out America," and "Best Friends." These pricing schemes give customers additional savings of 10% to 50% over the regular long-distance rates.

Divestiture promoted competition by bringing new players into the game. Competition between the different carriers helped to promote new technologies such as advances in switching technology and increased transmission rates. ATM,

Frame Relay, and ISDN are a small sampling of the new technologies available which provide business customers with increased bandwidth, transmission rates, and switching options. The Core Technologies section of this chapter will go into more detail about the technologies available today.

Customer Profile

There are enough telecommunication products and services offered today by the regional Bell holding companies, long-distance network carriers, and third-party vendors and resellers to satisfy the current needs and expectations of most individual consumers and businesses. In fact, there are so many communication options available that most people are not aware of them. Some options are not available to consumers in their residences because of physical limitations of the Plain Old Telephone Service (POTS) environment. Therefore, I will describe communication products and services from the point of view of two different environments: the POTS environment and the business environment.

Plain Old Telephone Service Environment

The telephone line obtained from your local telephone company provides you access to the public telephone network. That line is said to provide POTS (Plain Old Telephone Service), which is primarily used for residence service and single-line business service. POTS links homes and small businesses via copper wire and cable into the local exchange service. The local exchange service is provided by a public telephone center known as a "central office" within a designated geographical area known as the "exchange area" or "local service area." This center is usually owned and operated by your regional Bell operating company network. Calls made from one point to another in a particular exchange area are called local calls. In most metropolitan areas, calls to exchange areas that are near the local exchange are classified as "zone calls."

You may have seen, at one time or another, a commercial on television from some long-distance carriers boasting that their network is made up entirely of fiber-optic cable. You may have seen claims that the sound quality of these fiber optic networks is so good that you could hear a pin drop. The fact of the matter is that when you use a POTS line from a home or small business, you are limited by the channel capacity of the copper wire linking your home or business to the local switching center. This is true no matter which long-distance carrier you have selected.

Channel capacity is important because it describes the size of the communication path used for carrying information across a telephone line. For example, natural-sounding human speech is limited to a frequency range of 250 to 3,500 Hertz or cycles per second. POTS provides us with a bandwidth of 4,000 Hertz (4 KHZ), which is enough channel capacity to provide us with our basic communication needs. However, the bandwidth required to capture a color video signal is 4,000,000–6,000,000 Hertz (4,000 to 6,000 KHZ). This is why it would be nearly impossible to receive your television cable channels through the 4-KHZ channel POTS without some very complex compression technology. No commercially available technology today will achieve this level of compression. However, there are many labs in the United States currently working on developing this technology.

Due to the advances in digital switching in the central office or public telephone centers, many new features, products, and services are being offered with the basic POTS. Many of the value-added services available are provided by your regional Bell operating companies, long-distance carriers, and third-party Specialized Common Carriers. Value is said to be added to a network when additional services supplement the basic ability of establishing a simple connection between two subscribers. There are also many specialty hardware products available which can be used with POTS. Most of the features, services, and products available in the POTS environment are also available in the business environment, described later.

Many telephone service features available to POTS users are provided by your local telephone company. The following services described may be available or not, based on the equipment and switching system used by the local exchange center and the actual telephone instrument being used. Three-way calling service permits a customer to add a third party to an existing conversation. Call Waiting allows a call to a busy telephone to be placed on hold while an audible tone interrupts the called party as notification that another call is waiting. Call Forwarding enables a call to be automatically routed to another telephone number after a predetermined number of rings. Speed calling or speed dialing allows a caller to reach frequently called numbers by using an abbreviated telephone code instead of a conventional telephone number.

Some additional telephone service features which may be provided by your local telephone company are: Answer Call, which is a voice mail service similar to the one you may be accustomed to in most businesses; Identification Ring, which will identify up to six different numbers for you by ringing your telephone with two short rings repeatedly; Return Call, which will redial the telephone number of the last person who called you; Tone Block, which will temporarily prevent another call from interrupting your telephone session if you have the Call Waiting service. There are many other service features which may be available from your local telephone company. Your local telephone business office will say what is available in your area.

Another value-added service available to POTS is the electronic yellow pages. This service is provided by some of the local telephone companies. The service is available at no additional cost and allows access to a voice-automated response system which is loaded with prerecorded messages arranged in a variety of categories. The categories include information such as sports, news, reviews and previews, general interest, horoscopes, and the weather. This service is a type of voice-mail bulletin board. When telephone subscribers call in to a specific number and punch in a specific code, they are actually accessing a computer, which then routes the caller to a specific voice mail module that reads the recorded message. The telephone number and specific codes for each category are provided in your local telephone book yellow pages.

Long-distance providers, such as AT&T, MCI, Sprint, and others, also provide additional services to POTS users beyond the typical Direct-Distance Dialing (DDD) service, which permits you to dial outside of your local service area without the aid of an operator. There are many operator-assisted call services you probably have already used such as person-to-person calls, collect calls, and the calling-card service. There is also a third number call service that enables a long-distance call to be billed to an authorized third-party telephone number. Let's not forget the 1-800-COLLECT, offered by MCI, and 1-800-CALL ATT, offered by AT&T, which allows you to call someone collect by using a specific carrier. There is also a TIME and CHARGES service which allows you to obtain the length and cost of any call by asking an operator.

Other value-added services available to POTS customers are messaging services provided through packet-switched networks owned by long-distance carriers such as AT&T, MCI, Sprint, and other Specialized Common Carriers such as IBM and GE. Access to these networks allows you to send store-and-forward messages such as faxes, electronic texts, and binary file attachments (Lotus spreadsheets, graphic files, etc.), to one recipient or a large list of recipients, with nothing more than an ordinary computer, a regular modem, and some type of communication software. These networks also provide basic interactive access to proprietary databases and public databases. Belonging to one of these packet-switched networks does not limit your ability to communicate with people subscribed to other networks. You will be able to send electronic mail messages to people subscribing to other packet-switched networks by utilizing an addressing standard called X.400. I strongly recommend that everyone learn and adopt the X.400 international standard of addressing in order to communicate with the rest of the world.

Electronic bulletin board and database services are also available to POTS customers from Specialized Common Carriers such as CompuServe, Prodigy, and America On-Line.

Using the same computer, modem, and communication software mentioned above, you can access these database services interactively, perform searches, upload and download information, and chat with other subscribers. These networks were designed specifically for the individual consumer by providing services such as the ability to perform banking transactions, request sales literature, make credit card purchases, schedule airline reservations, play games interactively with other members or just chat on-line with other members interactively. Some of these networks also offer the capability to send faxes or electronic mail.

The Internet is yet another communication option available to POTS customers which is quickly gaining acceptance among consumers and businesses worldwide. It is neither a commercial network such as the ones owned by long-distance carriers nor is it a private network such as the ones owned by Specialized Common Carriers. The best way to view the Internet is as a conglomeration of thousands of Local Area Networks and stand-alone PCs connected together via packet-switching technology TCP/IP. A computer or terminal connected into the Internet can send/receive messages to/from any other computer on the Internet and can retrieve data from any electronic bulletin board attached to the Internet.

There are some hardware products designed to work with POTS which are available from regional Bell holding companies, long-distance carriers, and third-party vendors. Some of these products you may already be familiar with, such as the call-screening device, which displays the telephone number of the person who is calling you, or the various acoustic attachments which permit you to record a conversation you are having on the phone. One of the newer "high-tech" devices available today is the videophone. The videophone allows you to view the person you are calling if (and only if) that person also happens to own a videophone. This person's image on the screen of your videophone will not be smooth like the images on your television screen. You will visually notice fragmented motion whenever the person you are speaking to makes any movement. This problem is caused by the channel capacity of 4 KHZ of POTS, which does not allow enough compressed information to pass through the copper wire connected to your phone.

The Business Environment

If your business can afford it, there are hundred of options available to suit the requirements of your business and applications. It is important to select a long-distance carrier based on the long-term strategy and growth of your company. Long-distance telecommunication services available to the business environment are provided by a number of domestic and foreign carriers on a highly competitive basis. The technology, pricing structure, and corporate strategy vary from one carrier to the next. It is also important to evaluate your organization's needs carefully in order to match these requirements with the many telephone hardware and software features available. Note that there are many more products and services available than just the ones mentioned below.

One of the most common long-distance services available to businesses is the Wide Area Telecommunication Service (WATS). WATS is a pricing mechanism that allows businesses to make long-distance voice or data calls at a bulk rate instead of on an individual basis. 800 Service is another common long-distance service which permits callers outside of the local calling area to reach your business on a toll-free basis. The costs associated with the 800 service are charged to your company and are based on distance, time of call, and duration of call. Most 800 numbers are now portable among different carriers, which means that companies can keep certain 800 numbers even if they switch long-distance carriers. 900 Service is yet another long-distance service businesses can use to allow customers to call in for a specific purpose, such as registering a vote, while making the caller pay for the call. The 900 Service offered by long-distance carriers is not being widely used or accepted because of a negative, perceived association with fraudulent business activities and pornographic voice services.

Another common long-distance service available is the Foreign Exchange Service (FX). FX is a service which permits customers and employees to avoid long-distance charges whenever they are calling to or from a specific foreign exchange area. For example, if your company is located in Washington, DC, and you have a Foreign Exchange Service in the city of Philadelphia, PA, then customers and employees can actually avoid long-distance charges whenever they call to or from either telephone exchange. This is accomplished by your company renting a leased line from the long-distance carrier connecting your company's telephone switch to a central office (local telephone provider) in another geographical exchange area. Your company's telephone number could be listed in the telephone directory of the foreign exchange, which would mean, in the example I provided, that customers from the city of Philadelphia, PA, could call your company as if it were a local call. This service can provide your company with marketing advantages, such as portraying a local presence in a city before committing any resources to the area—allowing your products or services to test the waters.

Private-Line Service is another common long-distance service, which provides your company with the exclusive use of a leased line between two specific points. Based on the bandwidth of this private line, you could transfer voice, data, video, or any other type of transmission since you are bypassing the local telephone company and POTS. This type of service is commonly used between branch offices of companies that may have multiple locations throughout the country. There are many flavors of this service being offered by long-distance carriers such as ATM, Frame Relay, and ISDN, as well as many bandwidths available, such as 56 kbs and T1. These different technologies will be described in the Core Technologies section of this chapter.

You may have heard the term "PBX" mentioned before and wondered if it was the brand name of some telephone equipment your company owns. You are almost correct. PBX is a term which stands for a private internal telephone system. Local telephone equipment suppliers, specialty vendors, and long-distance carriers all provide different PBX equipment with an array of features and prices. Many of the features described should be available from most PBX equipment such as Automatic Call Back, which permits the caller to instruct the "busy" telephone to call back as soon as the "busy" telephone is free. Trunk Prioritization enables someone to use a WATS line or other service to the fullest by automatically stacking the calls of other users with lower priority in order to free up bandwidth. Automatic Route Selection (ARS), also called Least Cost Routing (LCR) provides the automatic selection of the most efficient and inexpensive route for calls originating from your business. Least Cost Routing usually selects a private leased line first if one is available, then it will select a WATS line as a second choice, then a Direct-Distance Dialing (DDD) line if all else fails.

Another PBX feature is Remote Access, which enables authorized employees to call from locations outside your company and place a security code to access your PBX system. Voice Mail or Voice Storage & Retrieval (VSR) is a computer application connected to the PBX which records voice messages digitally, stores them in a database, and forwards the message to the designated employee after he or she enters an ID and password. Voice Teleconferencing is another feature that allows employees at a number of different branches and facilities to tie into a conference bridge and converse among themselves. The bridge is an electronic device located in your PBX, at the local telephone company's central office, or as a value-added service from a long-distance carrier, which allows from three to hundreds of people to converse from different locations.

Teleconferencing provides major cost savings for businesses in terms of saving employee travel time and expenses. There are different methods of bridging your parties to a conference call. One way of bridging is to dial a specific access number, type in codes to signify the kind of call (audio, graphic, or audio-graphic), then dial the other parties into the call. Another more common way of bridging a conference call is to call the teleconferencing service ahead of time and reserve, for a particular date and time, the number of lines you will need to connect all parties to the call. You will then receive from an operator a specific access number and conference security code that each member in your conference call will need to use when they call in at the scheduled time.

Video-teleconferencing allows images of the attendees to be transmitted to the various conference locations by using sophisticated equipment and access lines. There are different quality levels available to display the video images. Some video conferences provide slow-scan video which changes the video image at specific time intervals, while others provide full-motion video. Sophisticated digital video teleconferences provide the ability to display multiple sites or images on the screen at the same time. Once again, the greatest problem limiting this technology is the available bandwidth connecting each of the sites and the maximum throughput based on the network architecture and topology.

Computer-conferencing services are yet another form of sharing information interactively with multiple attendees. Each participant uses their ordinary computer, modem, and communications software to dial into a central computer located on the service provider's network. Information, in the form of ASCII text, is then typed into a participant's computer and then transmitted to the central computer. The information is then accessible to all participants for as long as the computer conference call lasts. One advantage to this form of teleconferencing is that if a participant is late in joining the meeting, he or she can review on their computer screen all information typed since the beginning of the call. Another advantage to this form of teleconference is that participants can receive and add comments at their convenience.

Computer teletraining is yet another form of teleconferencing, which allows instructors to teach students, equipped with ordinary computers, modems, and communications software, at various sites across the country. The way this service works is that whatever an instructor writes on a digitized blackboard, tablet, or PC, is then broadcasted to each of the students' computers at each of the remote locations. Combine this service with regular audio teleconferencing with a speaker phone at each remote location and your training seminar now becomes one large classroom.

Electronic Data Interchange (EDI) is a value-added messaging service which allows companies to save time and money by exchanging computer-generated electronic documents in industry-specific standard formats. For example, a lot of time and cost is associated with manually generating a purchase order and processing invoices. It also takes time and money to send each document over the postal mail network. If a PO or invoice sent from another company does not include an important field you usually keep for your records, then it will affect your auditing trail and may slow the entire process. EDI changes this process by automating the generation of documents such as POs and invoices into industry-specific standards called transaction sets. These transaction sets are then almost instantaneously uploaded to a value-added messaging service or a long-distance carrier, which can deliver the document electronically to the recipient as an EDI transaction. If the recipient does not want to receive the document as an EDI transaction, the value-added messaging service can convert the EDI document into human readable form and deliver it via fax, electronic mail, or postal mail.

Distribution Systems

Telecommunication networks are designed to carry voice, data, or image transmissions from either a local or global geographical area. There are many ways to transfer this information between the local telephone companies, bypass carriers, and long-distance carriers. A communication session could take different paths depending upon the distance and location between the sending and receiving points. Information could travel through any number of transmission channels such as open wire, paired cable, coaxial cable, radio, satellite, waveguide, or optical fibers. Each transmission channel carries a different number of telephone or data signals. Sometimes, the transmission channel carries only a single signal, while others could carry many signals combined together (multiplexed). This section will describe the different networks available, their wiring topology, and the benefits or limitations of each topology.

The telecommunication industry in North America has a unique distribution system made up of local telephone companies known as Regional Bell Operating Companies (RBOCs), By-Pass Carriers, and Long-distance Carriers. Each RBOC has a defined LATA (local access transport area) that they must comply with under the 1982 divestiture plan. The consumer benefited from the divestiture because of the addition of new services and products, and AT&T would no longer control all the cross-connect fees. Consumers and businesses were left with a greater choice of services and products at substantial savings. Many competitors began to emerge, offering two basic types of networks known as public-switched networks (PSNs) and private-line voice and data networks. Public-switched networks can also be called Message Toll Service (MTS), switched network, long-distance network, interexchange facilities, and direct-distance dialing services. Private-line networks can also be called leased line services, bypass services, tie-line services, dedicated line services, and full-time circuits. But no matter what they are called, they fall into two main categories: public-switched and private-line networks.

Public-switched networks provide telephone services for voice and data transmissions to the business and residential community. Everyone shares common switching equipment and channels. Fees for this type of network are paid on a per-call, per-mile, per-minute basis. The FCC insures that each RBOC provides the same interconnect privileges to all Specialized Common Carriers (SCCs), such as all the long-distance carriers and Specialized Common Carriers. Public-switched networks provide POTS users with access to long-distance providers and Specialized Common Carriers.

A public packet-switching network is one example of a public-switched network which provides a more efficient method of transferring data. During packet-switching transmission, a message is divided into small units called "packets," which are then transferred individually over the network. Each packet contains specific information which reassembles the message in the proper sequence before it reaches its destination. The advantage of packet switching is that it uses the network for only a brief burst while in transit, versus a typical circuit-switched message, which utilizes a line channel for the entire duration of the message. This is a critical feature when you have a business environment with many users sharing data over the same line facilities.

Integrated Services Digital Network (ISDN) is an emerging international standard for a digital public-switched network. ISDN will integrate both voice and data at the same time over the same connection to the local telephone exchange. This means that it will no longer matter what kind of information is being transmitted over the public network: the user interface and protocol will be the same for all. This technology would enable someone to have a telephone conversation while transmitting data, such as video, binary files, or whatever, over the same transmission line. ISDN is explained in more detail in the Core Technologies section of this chapter. The advantages of ISDN for multimedia are obvious. Companies wishing to enter the multimedia industry should unite and push the ISDN issue at a national level through the FCC, the local RBOCs, and the long-distance and Specialized Common Carriers. The central office of each regional exchange area (controlled by the RBOCs) need to upgrade their systems to include ISDN standards.

Private networks or leased facilities are transmission lines and equipment that are dedicated to the exclusive use of a particular business or individual. Flat monthly fees are usually paid for this type of network by businesses or individuals who are using the particular "lines" on an unlimited basis. Private lines offer many advantages over public-switched lines, such as availability; there are no busy signals if you own the line (unless, of course, you did not lease enough lines for your company and your users overload the system). Another advantage is that private lines can be conditioned to improve transmission quality by reducing distortion and increasing transmission speed. Conditioning of lines requires use of electronic components to the circuit being used. The cost effectiveness of private lines over public-switched lines depend on the amount of usage since the flat monthly fee will be charged whether anyone has used the line or not.

Another form of private network is the Virtual Private Network, which is a mixture of public and private network circuits arranged in a customized fashion for a particular business. Virtual Networks are created with customized software within the long-distance carriers' telephone exchange, which provide enhanced network routing and management capabilities combined with the sharing of line facilities from their public-switched network. The main advantage of this type of network is that companies can enjoy the bulk pricing advantages of a private network with the efficiencies of scale of a large public network. The disadvantage of this type of network is that your company will usually need to make a minimum volume commitment over a particular time frame under specific terms of a contract with a particular vendor. Virtual networks are provided by all major long-distance carriers. AT&T calls their service "Software Defined Network (SDN), MCI calls their service "Virtual Networks" (V-Net), and US Sprint calls its service Vertical Private Network (VPN).

The high cost of interconnect fees between carriers and the marketplace caused a need for an alternative to the RBOC'S control of the local access. The development of "By-Pass" carriers has grown in recent years, providing a choice for many customers desiring lower rates and network resiliency. Some of the "By-Pass" carriers on the market today have built inner-city fiber rings that bring their service into the customer's facility. The By-Pass carrier then connects the customer's traffic to a long-distance carrier or routes it to the customer's different offices in the same city where they have established a point of presence. Other By-Pass carriers provide access through satellite network services, digital microwave services, private mobile radio services, and region-specific cellular services. The By-Pass market is growing, and each year more cities are added with the presence of these companies and services.

Long-Distance Network Design

The success of today's long-distance carriers can largely be attributed to their network design. The various transmission media available to the carriers provided the bandwidth and transmission speeds available in the current marketplace. The first carriers that entered the long-distance business soon realized that to keep up with market growth they needed to incorporate different media types that could provide capacity and be cost effective. The network would also have to be capable of interfacing with the existing POTS infrastructure and overseas telephone companies. The solution was building fiber networks where feasible as well as including other transmission media when the environment, such as mountainous regions, made fiber optics impractical. The long-distance carriers searched for rights-of-way in railroads, abandoned pipelines, Western Union underground facilities, and anywhere they could find them. Wiltel's president, Roy Wilkens, had the idea that existing pipelines not in use from the Williams Companies would make an ideal solution for installing fiber optic cables, and from that came one of the industry's largest specialty long-distance carriers.

During the early pioneer days of the telephone and telegraph industry, the Open-Wire pair cable was the only transmission means available. The Open-Wire pair consisted of two uninsulated, bare copper wires strung on wood poles, separated on glass insulators by a distance of about 12 inches to prevent short circuits during high winds or storms. This type of cable is subject to crosstalk (when the electrical signal transmitted in one pair leaks into another pair of cables) and other noise interference. Although the maintenance cost is high and impractical, you can still find this type of Open-Wire cable in rural areas of the United States. This technology eventually evolved into the Wire-Pair cable, which consists of two individually insulated copper wires twisted together with a full twist every 2 to 6 inches. This twisting reduced the noise interference between the pairs. The Wire-Pair cables were then tightly packed into larger cables, which could be either suspended on poles or buried in the ground.

Because of the high costs associated with placing Wire-Pair cables throughout the United States, telecommunication inventors quickly found a way to transmit more than one telephone signal over a single wire circuit. The sharing of one transmission medium by many signals is called "multiplexing." There are two forms of multiplexing being used today: analog and digital. Analog multiplexing, also called frequency-division multiplexing, divides the channel frequency range of a signal into a narrower frequency band. Analog modulation is achieved by the shifting of signals into a specific frequency through either the amplitude or frequency modulation of a carrier wave. Digital multiplexing, also called time-division multiplexing, assigns each digital signal its own unique time interval or time slot and then bursts a large number of these time slots over a particular transmission medium. Digital modulation is achieved by the sampling and encoding of an analog signal as a series of binary pulses.

Analog multiplexing is slowly being phased out by most long-distance carriers, although most transmission media today continue to receive analog modulation waves for light or radio signals. For example, thousands of voice and data channels may be digitally multiplexed together to go over a fiber optic cable; however, the actual light signal transmitted through the fiber optic cable is analog. The digitally multiplexed signal is used to turn on and off the analog wave of light being transmitted. The same holds true for microwave radio transmission. The digitally multiplexed signal must modulate an analog radio-frequency carrier wave that shifts the digital multiplex signal into the frequency band of the assigned microwave radio channel. Most transmission media continue to use analog waves, and that these analog waves represent signals that contain thousands of voice signals that have been digitally multiplexed together.

Coaxial cables was the next evolutionary step in transmission media. A coaxial cable consists of one or more hollow copper cylinders with a single wire conductor running down the center. The name "coaxial" comes from the fact that the cylinder and the center wire each have the same center axis.

Coaxial cable allows the baseband transmission of a digital signal in which the actual multiplex signal is transmitted as an electrical signal instead of an analog signal. A coaxial line can carry transmissions signals in only one direction because of the amplification required to transmit down the section of the cable. These amplifiers, called "repeaters," must be placed at a specific intervals along the cable or else the signals transmitted will become weaker and weaker and eventually disappear. A number of coaxial lines are placed together to form a coaxial cable for use in one- or two-way transmission systems.

The multiplexing system used with coaxial cables is called L-Carrier. As repeater and multiplexing technology improved, cost effectiveness improved and the overall capacity of coaxial cables increased. Coaxial cable is being replaced by optical fiber by all long-distance carriers. However, today coaxial cable is commonly being used for short-distance links in a specific geographical area. For instance, coaxial cable is most likely used to connect your television set to the local cable television service offered in your area. Another example of coaxial use is in Ethernet, which has become a Local Area Network (LAN) standard. Coaxial cable is ideal for distributing analog video signals to industrial and home receivers from terminating high-speed digital circuits. Coaxial cable is rather simple to use because most types of electrical signal can be directly connected. The bandwidth limitation of coaxial cable, however, places its digital capacity at a fraction of one optic fiber line.

A microwave terrestrial radio system is another commonly used transmission medium available today. A microwave radio system sends signals through the atmosphere between towers usually spaced at 20- to 30-mile intervals. Each tower in the system amplifies the signals and retransmits them at each receiving station until they reach the final destination. The signals follow a line-of-sight (straight-line) path and the antennas must be within sight of each other. The microwave radio emits extremely high-frequency waves in the gigahertz range. The radio waves are conducted from the antennas through metal pipes called "waveguides." Moisture and temperature can cause the radio beam emitted from the waveguides to bend, causing the signal to fade. However, the main advantage in using this medium is that you can carry thousands of voice channels, using digital time-division multiplexing, without the use of physically connected cables. This is a great cost-saving technology when trying to span mountains or heavily wooded areas, which pose great cost and physical barriers to wire and cable installations. Microwave terrestrial radio used to be the backbone of the long-distance network, but carriers are also replacing this medium with digital optical fiber because of bandwidth considerations.

Communication satellites are a transmission medium very similar to microwave terrestrial radio systems. In fact, the satellite functions as a microwave tower up in the sky. Solar-powered satellites orbit directly over the earth at a distance of 22,300 miles so that they travel at exactly the same speed as the rotation of the earth. Because of their altitude, they can pick up a radio beam from anywhere in the country and reflect it back to another portion of the world. Satellites can handle very large throughput of voice, data, and video transmissions simultaneously. Electronic circuitry on the satellite receives signals transmitted from the earth station. These signals are very weak and must be amplified by the low-noise amplifiers (LNAs) aboard the satellite. The "transponder," the circuitry used on the satellite, receives the microwave signal, changes its frequency so that the outgoing signal does not interfere with incoming signal, amplifies the signal, and then retransmits it. Most communication satellites operating in the most popular bandwidth typically carry 24 transponders because each radio channel on the satellite must have its own transponder. A single transponder or channel can be used for one full-color television signal, 1200 voice circuits, or digital data throughput up to 50 Mbps.

There are some inherent problems in using satellite technology as your transmission medium. For instance, the approximate time required to send a satellite radio wave from one earth station to another is about 270 milliseconds (ms) because of the distance between the satellite and earth stations. In a typical telephone conversation, one person speaks, then the other person responds. Well, under this scenario, it would take 270 ms for the initial speech to travel from one person to the other, then another 270 ms for the response to travel back. Therefore, from the time the first person finished speaking, it would take 540 ms (a little over a half-second) before he or she would hear a response—that is, if it took only one satellite hop to reach the destination. If someone in the United States was trying to reach someone in India, it might take two or three satellite hops to transmit the data. Under that scenario, you could be experiencing about a one- to one-and-a-half minute delay before hearing a response. Therefore, this transmission medium is best suited to the burst nature of data communications versus voice and video communications.

The most technological development made in transmission media is the introduction of fiber optics. Fiber-optic lines are hair-thin filaments of transparent glass or plastic that use light instead of electricity to transmit voice, video, or data. Covered to prevent light loss, the fibers are bundled together into a flexible cable. The optical fiber cables are then coupled with laser technology to generate very high-frequency beams of light with enormous information capacity. The transmission sequence begins with an electrical signal which is transformed into a light signal by a laser and fed into a glass fiber for transmission. The light signal is amplified along the way by repeaters. Then, at the destination, the light is sensed by a receiver and converted back into electricity.

There are many advantages in using fiber-optic systems as your transmission medium. For instance, fiber-optic cable is much smaller and lighter than any copper or coaxial counterpart. Therefore, it costs much less to manufacture and distribute fiber-optic cable than any other equivalent wire technology. A single fiber-optic cable with only two fibers can hold 1,300 two-way conversations simultaneously, while a traditional twisted wire pair will hold only 24 two-way conversations. In addition, fiber-optic cables will not be affected by electrical interference such as lightning or other types of electrical shocks. The advent of fiber-optic cable has revolutionized the long-distance telecommunications market and caused it to become the transmission medium of choice for most applications. The rate of fiber networks growth in the mid-1980s was phenomenal, with several small to medium-size telecommunication companies building their own networks. The market was changing so quickly that several of these smaller companies merged or were bought by larger companies wanting to expand their marketplace.

Right of Way (R.O.W.)

The issue of acquiring "right of way" for several long-distance carriers resulted in the leasing of spare fiber capacity on one or more fiber systems from direct competitors. There were also several areas where more than one carrier had the "light guide cable" in the same right of way as competitors. The disadvantage of this arrangement became evident when back-hoe operators working too close to the cables would often cut several cables at one time disrupting service for large numbers of customers. These early lessons of fiber breaks taught companies the need for true "network diversity." This helped the carriers' design networks that could route "around" the fault until it could be repaired.

The recent expansion of the U.S. markets globally resulted in new fiber-optic cables laid across the Atlantic and Pacific oceans. A consortium of international telecommunications carriers combined their resources to lay the cables across the oceans. These cables provided the needed bandwidth and transmission speeds required for multinational businesses operating large data networks. The installation of these fiber systems spanning such great distances required ingenious engineering because of the inherent problem, regenerating the lightwave signal. The addition of these new cables gives the long-distance carriers the ability to reroute traffic with a higher level of efficiency.

Core Technologies

The modern era of data and voice communications is rapidly changing with the emergence of new products and services offered by long-distance carriers, regional Bell operating companies, and specialty vendors. Consumers and business executives must be knowledgeable about the core technologies available in order to choose the products that best satisfy their applications. In this section, I will describe some of the technology available and list the advantages and disadvantages of using this technology.

ISDN

The concept behind Integrated Services Digital Network (ISDN) is to provide an all-digital network capable of integrating digital voice, data, digitized video, and digital facsimile signals over the same transmission and switching facilities. The ultimate goal of ISDN is to provide worldwide connectivity to private and public networks through commonly accepted international standards in a total end-to-end digital environment. ISDN is not proprietary to any particular vendor. Its interfaces and protocols are specified by the Consultative Committee for International Telephone and Telegraph (CCITT) under a United Nations division called International Telecommunications Union (ITU) that has 150 member nations. ISDN's system architecture follows the Open System Interconnection model (OSI), which was developed by the International Standards Organization (ISO). The OSI architecture is a set of software programs working together under seven layers or modules.

There are two configurations available today in the ISDN world. The first configuration, called "Basic Rate Interface" (BRI) or narrow-band ISDN, includes two B channels each operating at 64 Kilobits per second (Kbps) and one D channel operating at 16 Kbps. The B channels could be used for digital voice, video, and facsimile, as well as high-speed data communications. The D channel could be used for signaling and low-speed data communications. Telephone, fax, and video instruments need to have a built-in analog-to-digital converter in order to connect to the basic access interface. Several companies are now selling computer BRI interface cards for PCs and LANs to access ISDN services. The second ISDN configuration available, called "Primary Rate Interface" (PRI) or 23B+D primary rate, consists of 23 B channels, each operating at 64 Kbps, and a single D channel operating at 64 Kbps. Twenty-four 64-Kbps digital circuits multiplexed together is the standard DS1 (classic T1) digital signal. Therefore, you could say that this ISDN configuration follows the U.S. standard DS1 signal at 1.544 Megabytes per second (Mbs). The European version of ISDN offers 30 B channels and 2 D channels as their PRI rate at 2.048 Mbps.

One of the big advantage of ISDN over current T1 technology is in the area of call-by-call setup. In a current T1 environment, you need to allocate each of the 24 channels available for specific purposes. For instance, you might have 5 channels for inbound 800 services, 4 channels for videoconferencing, 10 channels for switched data connections into local area networks, and 5 channels for virtual network services. The reason the channel division is specific is because you need to let the long-distance or Specialized Common Carrier know what is coming across each channel at any particular time. Most of these carriers offer equipment and software to control channel partitioning. This type of channel allocation is not necessary with ISDN Primary Rate Interface (PRI) because the D channel of the PRI informs the carrier's network what is coming over the B-channel circuits. Therefore, the channel chosen by any particular application can vary on a call-by-call basis, making better use of your entire capacity. There are many other advantages to ISDN, making it the ideal choice for applications requiring infrequent, intermittent, and variable-size file and electronic-mail transfers, LAN to LAN connections to remote sites, telephone and videoteleconferencing, and maintenance backups.

In 1992, the Corporation for Open Systems International (COS), the North American ISDN User Forum, the National Institute of Standards and Technology (NIST), long-distance carriers, RBOCs, and specialty vendors met to agree on a set of technical requirements to make available interoperable, all-digital networks with standard network interfaces and uniform service capabilities. The National ISDN 1 standards made it possible for manufacturers to produce products that work with all ISDN services regardless of the switch handling the calls. Furthermore, the National ISDN 1 standardized signaling protocols and the message transfer layer between the telephone switch and the equipment on the customer's premises. In 1993, the National ISDN 2 added PRI functionality and broadened the overall offering. These standard interfaces should help ISDN gain acceptance in the U.S. business community.

ISDN has not yet emerged as standard for digital service carriers in the United States. In Europe, particularly in France and Germany, ISDN is widely available, relatively inexpensive, and there are many vendors selling BRI and PRI interface cards. ISDN has not done so well in the United States for many reasons. The primary reason that ISDN is not easily accessible is because there are approximately 15,000 local central offices (regional switching centers) which need to be upgraded to ISDN technology. Another reason for the slow acceptance of ISDN is that there are many digital service technologies available which are competing against ISDN, such as Frame Relay, Switched Multimegabit Data Services (SMDS), and ATM technologies. But ISDN is here to stay because the costs of BRI interfaces and services are continuing to decrease and the regional central offices are deploying more and more ISDN-compatible switches. One recent event which significantly helped the popularity of ISDN is the announcement from Microsoft Corporation that it will use BRI-ISDN technology to connect about 2,000 remote users to their corporate Ethernet LAN. In addition to data transmissions, Microsoft Corporation is planning to use ISDN for videoconferencing and group IV fax applications.

Frame Relay

Frame relay is a packet-switched data transmission technology which allows dynamic allocation of bandwidth between your facility and a remote site such as a Local Area Network (LAN). Frame Relay is targeted to large business networks by offering packet switching from 56 Kbps to 1.536 Mbs. Because the bandwidth is dynamically allocated, Frame Relay provides maximum utilization of network resources and equipment. Frame Relay will soon be offered at speeds up to 44 Mbs. There are also plans to map Frame Relay services into ISDN transmission capabilities. Frame Relay offers greater efficiency and bandwidth than X.25, the most commonly used packet-switched protocol. Frame Relay is faster than X.25 and is less expensive than a leased-line network. Because of this, more and more large businesses are using Frame Relay to connect their data networks (LANs) nationwide. One disadvantage of Frame Relay is that it has a variable-length delay which can degrade the quality of voice, music, or video transmissions. The best feature of Frame Relay is its capability to handle large "bursts" of data.

Switched MultiMegabit Data Service (SMDS)

Another technology well suited to data networks is Switched MultiMegabit Data Service. SMDS is a high-speed, cell-switched data service that allows businesses to move very large amounts of data simultaneously among many locations. SMDS works best with high-speed data applications in the Local Area Network (LAN) and Wide Area Network (WAN) environment. This technology supports transmission rates from 1.17 Mbs on 1.5 Mbs circuits to 34 Mbs on 44 Mbs circuits, with the potential to reach as much as 155 Mbs.

Asynchronous Transfer Mode (ATM)

Asynchronous Transfer Mode (ATM) is a standard cell-switching technique which allows users to change sustained information rates (minimum guaranteed throughput of a circuit) based on the time of day and day of week. Companies using ATM will be able to adjust the amount of bandwidth they are receiving at any time, thereby reducing costs to a minimum. The biggest advantage in ATM-based switches for private and public networks (including RBOC central offices) is that they will support nearly all the popular packet- and cell-switched services, including Frame Relay and SMDS. ATM technology integrates voice, data, and video traffic well because it can adjust the bandwidth to fit the application at any given time. The only problem with ATM is that it may take some time before long-distance carriers upgrade their switches to enable ATM technology to handle voice, data, and video traffic. The reason for this is that the big three long-distance carriers may lose revenue if their customers abandon their private or virtually private voice networks in favor of ATM. However, if the big three drag their feet too long, they may lose business to Specialized Common Carriers, who will first sell ATM for data applications, then turn around and sell voice and video services at lower rates than the larger carriers.

SONET

When fiber-optic technology was first made available, most vendors had systems that were unable to connect with other fiber-optic systems. The only way to connect two different fiber-optic systems was to demultiplex the optical signal, and then multiplex it again once the two optical fibers joined. The way to get around this was to develop an interface standard which could operate synchronously so that the transmitter and receiver could be locked together at a fixed rate of transmission. This interface standard is called SONET, which stands for Synchronous Optical Network. The basic SONET rate in the United States is 51.84 Mbps, while the SONET rate in Europe is 34 Mbps.

Common Channel Signaling System 7 (CCSS7)

Telephone service, whether it is regional or long-distance, involves a large network of transmission lines and switching facilities that need to be controlled. Each time you pick up the phone, signals are sent over the telephone network to control its operation, enable billing, and indicate status information such as that the line you are trying to call is busy. Interoffice "signaling" deals with those control and status signals sent between regional and/or long-distance switching systems. Until 1976, interoffice signaling occurred over the same trunk lines which carried voice signals. To reduce fraud and accidental disconnection, AT&T introduced a new signaling scheme in 1976 called Common Channel Interoffice Signaling (CCIS) which created a separate circuit between switching offices dedicated to the task of sending signaling information. This separate circuit connecting each switching office is called a CCS data trunk.

The switching equipment used in most central offices (switching centers) are controlled by digital computers and processors. CCIS allows two or more central office processors to digitally exchange information such as voice trunk availability, address information, and billing information. CCIS technology helped long-distance carriers use their voice trunk lines more efficiently. For instance, prior to 1976, when a long-distance busy tone was generated at a distant switching center (central office), the tone was sent all the way back to the central office near the calling party. This process tied up a full long-distance voice circuit. With CCIS, a data signal is sent back from the distant switching center over the CCS data trunk, and the busy tone is generated at the central office nearest the calling party. In this case, the voice trunks or circuits are free for actual conversation.

The signaling trunks on most long-distance and regional central offices are currently being upgraded to 56 Kbps digital circuits between the switching offices. This enhanced signaling system is called Common Channel Signaling System 7. Outside of the United States this enhanced signaling system is called CCITT 17. The OSI reference model was used to construct CCSS7. This signaling system supports important ISDN features, time-of-day routing, intelligent automated routing schemes, and Software Defined Network (SDN) features. Instructions for network routing and other important network information are stored in large databases that can be accessed, with CCSS7, by the equipment processors at the switching centers from anywhere on the network. CCSS7 is crucial to the development of worldwide digital networks because it provides central offices with the ability to manage and route new digital transmission signals as well as the common analog voice signals.

Video

The real-time dual transmission of voice and video images between two or more locations is probably a common dream among executives wanting instant feedback on decision making or strategic planning sessions without requiring out-of-town travel. In the past, cost was a major barrier for most businesses to enter the world of videoconferencing. Most corporate multipoint videoconferencing occurred over private DS1 (T1) lines among key locations. Most small to medium-sized business, health-care professionals, and educators could not afford the cost of private-line networks. The technology remained in the hands of the few since many applications required bandwidths up to 1.536 Mbs for high-quality images such as CAD/CAM images, x-ray imaging, and color photographs. Today, however, videoconferencing is gaining popularity because of lower costs and improvements in transmission technology.

In December 1990, an interoperability standard known as H.261 was endorsed by the International Telegraph and Telephone Consultative Committee (CCITT). The H.261 standard established that video bit signals should be sent across a network at "n" times 64 Kbps rates. Northern Telecom introduced a videoconferencing ISDN-based service called Dialable Wideband Service (DWS) in 1992. DWS is a circuit-switched bandwidth-on-demand public network that uses Primary Rate Interface (PRI) ISDN at the customer site and CCSS7 interoffice trunking at the central office switching center. These two technologies placed together offer customers dialable "n" times 64 Kbps (up to 1.536 Mbs) connections through the public network. With ISDN D-channel signaling, customers will dial the phone number and select the bandwidth desired (128 Kbps to 1.536 Mbs) on a call-by-call basis. This will provide state-of-the-art videoconferencing capabilities to anyone on an as-needed basis, without forcing customers to lease expensive private lines.

Network Management Systems

Current Network Management Systems have the resiliency to reroute network traffic automatically during service interruptions through the use of a "smart multiplexer." One example of the "smart multiplexer" technology is a product by Network Equipment Technologies (N.E.T.) called IDNX. IDNX automatically monitors the available bandwidth of a telephone network to enable service interruptions to be rerouted instantly. The product identifies the location of a fault on a private or public network by using a proprietary signaling channel using the overhead of a voice trunk/circuit. It then determines where the call can be rerouted and bypasses the fault with minimum impact to the end-user. The speed at which the call reroutes is often in milliseconds and the data stream that was being transmitted continues without experiencing any loss of information. This product, and other products similar to it (such as Newbridge), are currently being used to monitor networks in many U.S. and European companies.

Long-distance fiber optic networks have excellent network management tools available to perform monitoring and performance diagnostics. These tools include color-coded graphical views of the network, which change from one color to another depending on the severity of a fault. The graphics can also be expanded to represent switching centers. This feature provides network technicians with the ability to decrease the total time of repair normally needed when a service fault occurs. Other network performance tools allow instant analysis of service availability by checking each circuit on a network and by calculating the percentage of error-free transmissions. This performance tool can be set to show a comparative view of the different long-distance carriers used on the network and determine which carrier is maintaining the highest availability figures. This feature can easily be implemented and is an excellent choice for networks that are built around T1s.

Applicable Standards

New technologies are being developed and/or improved daily by hundreds, if not thousands, of communication carriers, manufacturers, suppliers, and vendors throughout the world. How do all these products and services fit together to allow for worldwide communication? In the past, proprietary interfaces, converters, and adapters were the common way to make everything fit. However, this solution is costly and ultimately affects performance. Clearly, interfaces and protocols must be standardized or else total chaos will result. The difficulty comes when multiple standards gain wide margins of user acceptance and these standards are virtually incompatible. For example, the International Organization for Standardization (ISO) created the Open Systems Interconnection (OSI), IBM created the Systems Network Architecture (SNA), and the Department of Defense (DOD) created the Transmission Control Protocol/Internet Protocol (TCP/IP). Each of those protocols evolved from a standard and gained acceptance from large groups of users.

A communication protocol is a predetermined and mutually agreed upon step-by-step procedure whereby both the sender and receiver understand how the communication session will begin, proceed, and end. Many protocols have been developed and implemented in the world of telecommunications. Some of these protocols have been accepted as national and/or international standards. Other protocols have been adopted, not by official standards, but through widespread usage. However, there is a substantial price to be paid for lack of standardization in communication networks. Proprietary protocols necessitate proprietary equipment and software. This reduces the options and flexibility of the customer and supplier. Also, nonstandard network equipment and interfaces add a level of complexity and costs, and have the effect of locking you in to a particular technology.

Open Systems Interconnection (OSI)

In 1977, in an attempt to standardize the many interfaces, protocols, and standards encountered in data communications, the Geneva-based International Organization for Standardization (ISO) created a conceptual model called the Open Systems Interconnection (OSI). The OSI reference model created rules which demand that every communication task remain in assigned layers and that the output of each layer precisely matches the format established for it. OSI is a distributed systems architecture standard which has gained international acceptance. The seven layers of the OSI model are:

1	Physical layer
2	Link layer
3	Network layer
4	Transport layer
5	Session layer
6	Presentation layer
7	Application layer

Before data are transmitted, each OSI layer in the sending device or equipment establishes the ground rules for the communication session with the corresponding layers in the receiving device or equipment. The first three layers (physical, link, and network layers) control the data transmission method through the communications network. The physical layer deals with the actual physical and electrical connectivity of the hardware. The link layer deals with the stream of bits called "frames." The beginning of a "frame" has a header, which contains address information, and a trailer, which indicates the end of a frame. Error detection and correction is also specified in the link layer. The network layer deals with the actual transmission and routing of the data through the network.

The four "higher" level layers (transport, session, presentation, and application layers) deal primarily with the software and application being used. The transport layer deals with how the data are handled when errors are detected and how to handle data overflows when the network is congested. The session layer deals with session-specific information such as user identification and recovery from transport failure. The presentation layer deals with the encryption of data, the positioning of the cursor on the display screen, the encoding of graphical symbols for display, and the standards needed to represent various types of data files. The application layer deals with the actual task or job that the user or software is currently working on.

The goal of the OSI reference model is to provide for geographically distributed processes the ability to implement applications cooperatively. Applications running under the OSI system architecture will be able to communicate without regard to differences in hardware, operating systems, or data representations. The seven layers of the OSI reference model should be used as a blueprint for developing your application software if you want your application to run across multiple platforms using an internationally accepted standard.

Systems Network Architecture (SNA)

IBM has its own proprietary layered data networking protocol standard called Systems Network Architecture (SNA). Introduced in 1974, the structured layer of SNA resembles that of the OSI model. SNA, however, does not provide full-scale support of the OSI protocols. The physical and data link layers of the SNA architecture are functionally similar to the two lower layers of OSI. The higher layers of SNA are considerably different from OSI due to the host orientation of IBM's architecture. For instance, OSI provides a network layer to handle the routing of packets from node to node and a transport layer to establish a connection for sending or receiving messages. SNA, on the other hand, utilizes a path control layer via the host to establish a connection in the form of a virtual circuit to a remote. The path control layer also routes packets between nodes and the host. In SNA, a session is an established connection via a virtual circuit from host to end-user device, allowing applications to be run on the host computer.

The differences between OSI and SNA become obvious when comparing the protocol layers of the two architectures:

OSI	SNA
Application	Applications Transaction services
Presentation	Presentation services

OSI	SNA
Session	Data flow
Transport	Transmission
Network	Path control
Data link	Data link
Physical	Physical

Transmission Control Protocol/ Internet Protocol (TCP/IP)

TCP/IP is a full set of protocols initially conceived to provide a level of hardware and software interoperability on the Defense Data Network (DDN) and other government-sponsored networks such as the Internet. TCP/IP has the largest vendor support base and user population, particularly among universities, military and other research laboratories, and private businesses connected to the Internet. Unfortunately, TCP/IP is not in compliance with ISO's OSI protocol standards. The intention of TCP/IP was to establish uniform data communication applications over disparate computer systems and communication network architectures. It did so by covering all the necessary layers required for effective data communications. TCP/IP is very well ingrained in many large networks. For instance, TCP/IP is integrated into network systems that employ UNIX as an operating system. TCP/IP has also been used extensively in local area networks (LANs) and wide area networks (WANs). The Internet, the most widely used packet-switched network in the world, uses TCP/IP as the standard protocol for communicating among its members.

The basic composition of TCP/IP comprises three protocol layers: the application layer, the transport layer, and the network layer. The application layer is designated the "upper layer protocols" (ULPs). One standard application running in the ULP is Telnet, a terminal-to-host communication software that allows PCs to access host computer systems over the TCP/IP structure. Another application running in the ULP is the file transfer protocol (FTP) which facilitates efficient and error-free transmission of bulk data files. The next layer in TCP/IP is the transport layer, for access into the data communications network. This layer, known as the transmission control protocol (TCP), can set up a virtual circuit over the network, connecting a ULP application with its counterpart on the network. This layer follows the concept of OSI; however, it is completely incompatible with OSI at this time. The last layer is the network layer called the "Internet protocol" (IP). IP transmits data in independent packets. This approach provides flexibility when dealing with different transmission modes and systems from various domestic and international networks. So the way these protocols work is that IP routes packets of information individually over the network, and the TCP layer above collects the packets, arranges them in correct order, and delivers them to the end-recipient's application, controlled by the ULP.

Infrastructure

Long-distance carriers, RBOCs, Specialized Common Carriers, vendors, and suppliers have all contributed to the building and maintenance of the copper and fiber optic network servicing most consumer residences and businesses today. Private networks can either tie into the public telephone network or bypass it completely with existing technologies. Consumers and businesses have inherited the current telecommunication infrastructure. However, the post-AT&T divestiture era has fueled competition, enabling new technologies to be developed and implemented. These new technologies have had an important impact on the telecommunication industry in two major areas: transmission and switching.

The implementation of fiber-optic systems has revolutionized the industry with regard to transmission capacity, transmission media size, and costs. This technology has significant implications for integrating distinct classes of transmissions, such as voice, data, video, and binary files such as CAD/CAM images and financial spreadsheets, with improved utilization of existing facilities and greater economies of scale. The enormous capacity of optical fiber combined with the reduced size and cost of the media make the technology the ideal choice for applications requiring digital transmission and large bandwidths.

Terrestrial and satellite microwave transmission technology has presented both advantages and disadvantages to the telephone network. The advantages include distance independence and relatively high capacity compared with local twisted-pair wire and coaxial cable. However, the disadvantages include signal delays due to the distances involved, inherent analog limitations, and vulnerability to interception and interference. Very small aperture terminal (VSAT) satellite earthstations, typically less than 7 feet in diameter, will provide an acceptable alternative to copper cables when sites are located in remote regions and when data rates are not crucial.

Switching technology evolved from a completely manual process, to an electromechanical physical connection, to a computer-controlled electronic physical connection, to a completely digital computer-based connection technology. This technology has reduced the cost and size of the switching facility, while dramatically increasing the capability and capacity of the switching equipment. Additionally, the digital nature of the new CCSS7 switching technology provides the necessary functionality for emerging transmission technologies such as integrated services digital network (ISDN).

The long-term commitment of long-distance carriers, specialized common carriers, and RBOCs toward fiber-optic technology is to provide a uniform telephone service, employing digital switches and transmission technology that are compatible with each other. The structure of this digital network is such that large amounts of spare capacity could be separated and leased to other organizations for private telecommunication needs. The same digital infrastructure exists in international telephone networks; however, the standard unit of digital capacity is very different. In the United States, the standard unit or building block of digital hierarchy is the T1 channel (DS1) with a capacity of 1.544 Mbs. The smallest standard unit in the T1 infrastructure is the DS0 channel, which provides 64 Kbps of capacity. The standard T1 carries 24 DS0s used to transmit digitized voice or data channels using 64 Kbps pulse code modulation (PCM). In Europe, the standard unit or building block of digital hierarchy is 2.048 Mbs, which is made up of 32 individual channels of 64 kbps each.

No matter how technologically advanced the transmission signals or switching centers become, communication networks will always be limited by the "last mile" or "local loop," linking the end-user's equipment to the first concentration point or regional central office. The costs associated with any changes to the local loop are much greater than any other section of the telephone network because this section of the network is dedicated to the use of a single user, home, or small business. Costs associated with switching centers or regional central office trunks are shared by the traffic of many callers. For this reason, do not expect to see many implementations of high-capacity links in the local loop until the costs of the new high-bandwidth technologies greatly decrease. Therefore, all the new high-tech digital services being discussed, such as television programs on demand, videoconferencing, and interactive virtual reality, will not occur over the public telephone network until a major investment is made in the current local loop, enabling end-to-end digital services. Other transmission media could provide tangible alternatives to the local loop limitation, such as coaxial cable, owned by various regional cable television companies. Another alternative to the local loop limitation is the radio spectrum provided by local cellular telephone networks.

Regulatory and Legal Constraints

The United States government found it necessary to regulate the telecommunications industry because of the belief that basic telephone services were considered essential for the public welfare. It seemed necessary for the government to control regular business practices in order to guarantee uninterrupted service and fair pricing. Since the telecommunication industry supplied services which were essential to the public good, the U.S. government classified the industry as a "natural monopoly." The reason AT&T grew into the largest monopoly ever allowed in this country stems from the government belief that only a single company could operate a nationwide telephone network efficiently as a natural monopoly. Encouraging competition in this industry would only create incompatible networks and a duplication of facilities and resources.

The principle of natural monopoly still applies today for the seven regional Bell holding companies which control and operate local telephone services of 22 local telephone companies across the United States. However, this monopolistic principle no longer holds true for long-distance telephone services since the January 8, 1982 "landmark" agreement between the U.S. Department of Justice and AT&T. On this date, AT&T agreed to give up its 22 local Bell operating companies in exchange for the Justice Department's dropping the antitrust suit levied against AT&T. District Court Judge Harold H. Greene reviewed the terms of the settlement and dismissed the antitrust suit on August 24, 1982. AT&T filed its divestiture plan in December 1982. The plan proposed grouping the 22 local Bell operating companies into seven independent regional corporations, each with its own board of directors and stock. The plan was approved on August 5, 1983, and the divestiture took effect on January 1, 1984.

The terms of the 1982 settlement between AT&T and the Justice Department changed the fabric of the telecommunications industry forever. It allows competition in the long-distance market and in the manufacturing of telephone equipment. AT&T agreed to divest all aspects of its business that provided local exchange services. They agreed that they would have no special relationship with the divested Bell operating companies (BOCs). They agreed that the divested BOCs would provide equal access to all long-distance network carriers. They agreed that the divested BOCs would not discriminate in favor of AT&T's telephone products and services. The divested BOCs would be allowed to provide local telephone services in specified Local Access and Transport Areas (LATAs), however, they would be prohibited from providing long-distance, information, or nonregulated (non-tariffed) services.

The principal cause for the 1982 divestiture agreement was that the U.S. government felt that AT&T was no longer acting in the best interest of the public good. The U.S. Justice Department felt that AT&T was subsidizing certain products and services in certain segments of the industry in order to ward off competitive products and services from companies such as MCI and Western Union. It was felt that AT&T was charging artificially high rates for some products and services while maintaining artificially low rates for other products and services. Therefore, in November 1974, the U.S. Department of Justice filed an antitrust suit which accused AT&T, Western Electric (the manufacturing division of AT&T), and Bell Laboratories of conspiring to "prevent, restrict, and eliminate competition from other telecommunications common carriers, private telecommunication systems, and manufacturers and suppliers of telecommunications equipment."

The suit asked that AT&T divest its manufacturing division, Western Electric, and either its long-distance telephone network or its 22 local Bell operating companies.

Current FCC Regulations and Legal Constraints

As part of the Modified Final Judgment (MFJ) requirement imposed by the Justice Department during the divestiture settlement, District Court Judge Harold Greene must review the ruling every three years and issue a report which will either amend or uphold regulations controlling the Bell operating companies. The first MFJ triennial review was issued in September 1987. The first review upheld the regulation prohibiting BOCs from entering long-distance services, manufacturing and enhanced services. However, in March 1988, Judge Greene amended his ruling and allowed BOCs to enter the voice and electronic mail services market. In July, 1991, during the second triennial review, Judge Greene changed his ruling once again and allowed BOCs to provide all forms of information services such as electronic publishing and cable television systems.

In March, 1989, the Federal Communications Commission (FCC), issued a price cap regulation for AT&T and many, but not all, Local Exchange Carriers (LECs) controlled by the BOCs. This regulation changed the way a telephone company could earn profits. Instead of a Rate of Return (ROR) regulation which restricted companies from earning profits above a certain rate of return, price caps regulation provides companies with a tariff "cap" that rates cannot exceed and a tariff "floor" beneath which rates cannot drop. The advantage of price caps is that they allow carriers to develop their own pricing method, including justification for a higher price if they can prove it is warranted.

Since the late 1960s, the FCC has been preempting state regulations in order to maintain greater competition in the telecommunication market. Preemptive FCC rulings such as the 1968 Carterfone decision, which forced telephone companies to allow telephone attachments and peripherals to interconnect to their networks, have been issued to prevent state regulation from prohibiting companies from providing federally approved telecommunication products and services. However, in 1986, the Supreme Court, in Louisiana Public Service Commission vs. FCC, ruled that the Communications Act of 1934 prevents the FCC from preempting state regulations. The Communications Act of 1934 created the FCC, responsible directly to Congress, and charged it with the regulation of interstate and international communications by radio, television, wire, and more recently, satellite and cable. The Act plainly states that the FCC is not responsible for regulating intrastate communications. This Supreme Court ruling has significantly weakened the preemptive powers of the FCC and will probably affect the manner in which new competitors, products, and services are introduced in the U.S. telecommunications market.

In 1992, the senate passed a bill, called the Cable Television Consumer Protection and Competition Act, which stated that local authorities could set basic cable rates provided they are in areas where there is no competition. The FCC, however, would regulate cable rates for non-basic service. In July 1992, the FCC ruled that phone companies could provide a video dial tone, permitting them to transmit cable television programs, and permitting the ownership of up to 5% of a cable programming company. In December 1992, Bell Atlantic filed a suit against the FCC and the U.S. Department of Justice to argue the constitutionality of the telephone/cable company cross-ownership prohibition in the Cable Communications Act of 1984. Bell Atlantic won the U.S. District court case in August, 1993, and purchased on October 13, 1993, two cable television networks: Tele-Communications Inc. (TCI) and Liberty Media Corp. The government could choose to appeal!

In 1991, Congress passed the Telephone Consumer Protection Act which allows the FCC to regulate the telemarketing industry. The FCC passed several regulatory measures which took effect in December, 1992. For instance, it is illegal to use autodialers and prerecorded voice messages to make calls to emergency, radio, health care facilities, and other services' telephone lines if the called party will incur a charge. The regulation also prohibits prerecorded messages to residential telephone numbers without prior consent and restricts the hours in which telemarketers can call a residential telephone number (between 8:00 A.M. and 9:00 P.M.). Businesses which fail to comply with these regulations could be subjected to costly legal action.

Outlook

We are living in exciting times. The telecommunications industry is changing so rapidly that it is becoming increasingly difficult to keep up with all the changes. In an effort to gain market share, expand services, and acquire new technology, a new trend is emerging, forcing companies to form alliances, mergers, or purchase other companies in order to remain competitive.

One recent global alliance was formed when British Telecom spent $4.3 billion in 1993 for a 20% stake in MCI Communications. In addition, the two companies have a total of $1 billion invested in a joint venture involving Syncordia Corp., an Atlanta-based BT-owned company which manages and integrates private corporate networks. This alliance was formed to compete with AT&T for the top 2,000 major multinational corporations, with worldwide operations, that are looking to "outsource" their corporate communications networks.

AT&T spent $12.6 billion in 1993 to purchase McCaw Cellular Communications, Inc. This purchase provides AT&T with 2.2 million cellular phone subscribers in 100 cities and approximately 20% of the U.S. market share. AT&T has also acquired or invested in a dozen other companies in the last five years. These companies provide various services or products such as manufacturing interactive multimedia equipment, computer equipment, advanced messaging services, modem equipment, personal communicator equipment, hand writing recognition software, wireless communication software, digital cordless phone equipment, video game software, on-line interactive entertainment networks, and multimedia educational software. The reason for these acquisitions and investments is that AT&T wants to provide the capability of delivering voice, electronic messages, faxes, and video to any device, anywhere, such as phones, computers, TVs, cellular handsets, and wireless devices such as the personal communicator. The strategy of wanting to "bypass" the local loop in the United States and in foreign countries could cause most foreign and domestic carriers and governments to try to block their efforts.

One recent failure in the merger "fever" was Bell Atlantic's purchase of Tele-Communications Inc. (TCI) and its sister cable programming company, Liberty Media Corp. for a reported $12 billion. Digital technology would have permitted these companies to offer new interactive products and services such as interactive games, home shopping, interactive instruction classes, and movies on demand. Bell Atlantic would have been able to expand into new markets by adding video to its telecommunications network and phone capabilities to its newly purchased cable network. The market share of Bell Atlantic would have increased to 22 million customers in 59 of the top 100 U.S. cities. This merger would have given Bell Atlantic a definite competitive advantage over the other 6 regional Bell operating companies. However, due to U.S. government regulatory constraints, the deal collapsed.

Direction of Products and Services in the Consumer Sector

When the next-generation technology is made available to the consumer market, it will revolutionize our lives at home and at work (which may end up being in the same location). Some of the following products and services already exist in the lab or test environment. However, the network infrastructure, which would provide the necessary bandwidth or channel capacity into our homes, cannot yet handle the amount of information required by the technology. It will take a lot of time, money, integration of different products and services, coordination of effort between different companies, and legal battles against local and federal legislators and FCC regulations.

Regional Bell operating companies (RBOCs) will continue to increase their value-added offerings to the consumer. They will play a major role in the explosion of information and entertainment services if they acquire cable companies which have high-capacity cable lines already installed in large customer base areas. So far, USWEST has a $2.5 billion stake in Time Warner Inc., the second largest cable provider. Other RBOCs will need to get in the game fast if they do not want to remain stuck with the slow-growth business of POTS. One of the new services you can expect will be the ability to send and receive phone calls and view the person you are speaking to on the screen of your television set. Another service will allow you to select a movie from a wide variety of database choices located on the telephone/cable network at any time of the day or night. There will be many on-line interactive services which will allow consumers to shop, make banking transactions, invest in the stock market, and play games with many other players. However, none of these technologies will be made available until the "local loop" limitations are dealt with.

Long-distance carriers will also increase their stake in the consumer market. They will provide products and services which will empower consumers to control how they want to interface with the network. People will be able to tell the network that they are ready to receive all calls, or that they want all calls to be held in voice mail until a later time, or that they only want to receive calls from this one person while all other calls get placed in voice mail, or that they want to be completely disconnected from the network and do not want anyone to reach them (when on vacation, for instance).

Expect people to have a unique telephone number that follows them wherever they go so that others could reach them in an emergency. This unique telephone number, assigned by the long-distance carriers, will serve as the single address ID for a person's telephone number, voice mailbox, video mailbox, fax mailbox, and electronic mail mailbox. So whether you are using your phone, television set, personal communicator, computer, or any other device, the network will inform you if you have any voice messages, electronic mail messages, video messages, or fax messages. You could then retrieve them in their original form or convert them to voice and listen to the message. Expect intelligent telecommunication devices that are capable of recognizing human voice commands, converting voice to text and text to voice. One day, you may hear someone on a bus saying the following to his personal communicator, "communicator, retrieve my electronic mail messages and read them to me."

Direction of Products and Services in the Business Sector

Technological changes in the telecommunications industry will affect major businesses in a very positive way. New products and services, offered by long-distance carriers, Specialized Common Carriers, and third-party service and manufacturing providers, will do much more than just improve the switching technology and increase transmission rates. Many products and services will be tailored for industry-specific applications to help companies improve on their bottom line. Communication vendors will need to become concerned with business issues such as improving cash flow, reducing inventories, reducing charge backs, and improving customer service, in order to compete in the business sector market.

More and more large businesses will decide to "outsource" their communication requirements to long-distance carriers. It is only a matter of time before these large companies realize that the expense in time, personnel, money, facilities, and equipment for building, operating, and maintaining large communication networks is not cost effective. It does not make sense for a company to spend the energy, time, and expense to try and keep up with all the technological changes and complexities unless it is in the business of selling communication services. Long-distance carriers will provide state-of-the-art facilities, hardware, software, and expertise at specific costs which can be measured and budgeted into a company's business plan.

Bypass technology will also gain popularity among most businesses in the years to come. Bypass technology involves developing ways of circumventing the local public carrier facilities (LECs owned by the RBOCs). Bypass technology is desirable if a company plans on using advanced technologies such as videoconferencing and digital high-speed facsimile. Selecting a Bypass carrier service provides businesses with a choice to select the best access to long-distance carriers for the lowest cost.

Summary of Multimedia Related Organizations

The following table lists many of the numerous organizations that are active to some degree in multimedia related businesses or technologies. Many of these groups are organized to represent vertical industry interests. Some provide individual membership activities while others serve primarily corporate interests. Still others operate as open standards groups. Since organizations change and evolve rapidly, contact them directly for up-to-date information.

Name	Full Name	Principal Membership	Description/Mission	Comments	Location	Contact Information
AIMIA	Australian Interactive Multimedia Industry Association	Individual & Corporate	Chartered to promote the growth of Multimedia in Australia and Southeast Asia and to coordinate with the Australian Government on multimedia related projects.	Relatively new organization	Scarborough, Western Australia, Australia	P: 61 9 383-4695 F: 61 9 383-2707
BIMA	British Interactive Multimedia Association	Corporate	Representative body for the British interactive multimedia industry. Covers all markets for multimedia.	Historically focused on Videodisc based applications for corporate training & sales	London	P: 44 733 245700 F: 44 733 240020
COS	Corporation for Open Systems	Corporate	Created to serve as the entity through which industry could convene to work the technical and programmatic issues that required common, agreed-to industry specifications.	Primarily U.S.-based Telecommunications service or equipment providers	Fairfax, VA.	P: 703 205-2760 F: 703 846-8590
DAVIC	Digital Audio Video Council	Corporate	Purpose is the promotion of the success of emerging digital audio-visual applications and services. Plans to accelerate the establishment of standards for video on demand type services.	Strong participation by international telecommunications service and equipment providers	Torino, ITALY	P: +39 11 228 6120 F: +39 11 228 6299
DSM-CC	Digital Storage Media Command and Control	Corporate	The DSM-CC Extension is specified in part 6 of the ISO/IEC 13818 (MPEG-2) standards. It specifies protocols for supporting multimedia applications, such as video-on-demand, on MPEG systems that are deployed in diverse and heterogeneous network environments.	This group is producing detailed and highly technical specifications for multimedia network protocols	(ISO - International)	Chairman: Chris Adams P: 408 944 6409

Name	Full Name	Principal Membership	Description/Mission	Comments	Location	Contact Information
EFF	Electronic Frontier Foundation	Corporate & Individual	Mission is to ensure that the new information infrastructure emerging from the convergence of telephone, cable, broadcast, and other communications technologies enhances free speech and privacy rights, and is open and accessible to all segments of society.	Strong populist Internet history and perspective; historically very active on policy issues on Capitol Hill.	Washington, DC	P: 202 347-5400
EIA	Electronic Industries Association	Corporate	Primary consumer electronics manufacturers and retailer trade group with strong policy and technical activities. Operates the CES exhibition. Has helped to develop a draft analog set-top decoder interface standard.	Has historically played key role in establishing television receiver standards together with NAB & FCC	Washington, DC	202 457-8700 202 457-4985
EMF	European Multimedia Forum	Corporate	A non-governmental organization, representing all parties involved in the multimedia community who share a common interest in the successful production, delivery, and use of multimedia technology and services in Europe.	Has strong support of the European Commission. Has similar agenda to US IMA but focused on Europe	Brussels, Belgium	P: +32 2 219 0305 F: +32 2 219 3215
ICIA	International Communications Industries Association	Corporate & Individual	Specialized Washington networking, professional training and development, annual convention, publications and information library, government relations, market research.	Founded in 1939; historical focus on audio visual communications	Fairfax, VA	P: 703 273-7200 F: 703 278-8082

Name	Full Name	Principal Membership	Description/Mission	Comments	Location	Contact Information
IDSA	Interactive Digital Software Association	Corporate	First organization dedicated to meeting the needs of the interactive entertainment industry. Represents interactive entertainment software publishers, including cartridge, video game CD, PC-CD, and floppy disk platforms.	Founding companies are mostly cartridge-based video game companies. Established in 1994.	Washington, DC	P: 202 833-4372 F: 202 833-4431
IEEE	The Institute of Electrical and Electronics Engineers	Individual	World's leading organization of electrical engineer professionals offering publications, meetings, and technical and educational activities for computer scientists and engineers. Promotes information exchange.	Professional society with strong technical and academic focus and representation	Washington, DC	P: 202 371-0101 F: 202 728-9614 "Information Line" recording 202 785-2180
IIA	Information Industry Association	Corporate & Individual	The IIA provides a forum for networking education and government relations activities supporting the needs of content, technology, and communications companies.	Strong focus on CD-ROM based delivery of information and data	Washington, DC	P: 202 639-8262 F: 202 638-4403
IICS	International Interactive Communications Society	Individual & Corporate	Professional organization of the technologies community. IICS provides a forum to share ideas, applications, and techniques for the effective use of interactive multimedia.	Caters to regional communities of interest through local chapters; majority of members participate as individuals	Beaverton, OR	P: 503 579-IICS F: 503 579-6272

Name	Full Name	Principal Membership	Description/Mission	Comments	Location	Contact Information
IMA	Interactive Multimedia Association	Corporate	Mission: To promote the development of interactive multimedia applications and reduce barriers to the widespread use of mm technology. Compatibility project, Intellectual property project, IMA Convergence Forum.	Strong technical and policy focus. Designed to provide a cross-industry forum with participation from various multimedia related industries.	Annapolis, MD	P: 410 626-1380 F: 410 263-0590 Fax info. (call from fax handset): 410 268 2100
IMDA	Interactive Multimedia Development Association (Canada)	Individual	To assist content developers in developments and marketing, provide a communications network, educate the public community, be a clearing house for information.	Strong (Canadian) multimedia developer focus.	Ontario, Canada	P: 418 325 8892 F: 418 325 8885
ISA	Interactive Services Association	Corporate	Only national organization solely devoted to the promotion and development of telecommunications-based interactive services in North America.	Strong on-line service orientation and interest	Silver Spring, MD	P: 301 495 4955 F: 301 495 4959
ITVA	International Television Association	Individual & Corporate	Mission is to serve the needs and interests of its members, to advance the video profession, and to promote the growth and quality of video and related media.	Video production and corporate video focus.	Irving, TX	P: 214 869-1112 F: 214 869-2980
MDG	Multimedia Developers Group	Individual & Corporate	The MDG's mission is to help emerging multimedia software companies become commercially viable by facilitating communication between the parties who develop, fund, service, sell, or regulate multimedia titles.	Membership focused in the San Francisco area; focus is on title developers (e.g., CD-ROM)	San Francisco, CA	P: 415 553-2300 F: 415 533-2403

Name	Full Name	Principal Membership	Description/Mission	Comments	Location	Contact Information
MMA	MIDI Manufactures Association	Corporate	The MMA is a trade association that represents hardware and software companies with an investment in MIDI technology. Provides a forum for discussion of implementation and enhancements to MIDI.	Creators and sole caretakers of the MIDI (Musical Instrument Digital Interface) standard and its derivatives.	La Habra, CA	P: 301 947-4569 F: 301 947-4569
MMA	(Japanese) MultiMedia Association	Corporate	Objectives are to establish an infrastructure for the multimedia industry, to promote its advanced usage and to create prosperous society with multimedia.	Japanese Government (MITI) supported group; significant software research focus.	Tokyo, Japan	P: 03 (3506) 1701-5 F: 03 (3506) 1739
MMCF	Multimedia Communications Forum	Corporate	MMCF is a non-profit organization devoted to accelerating multimedia network applications and to developing the recommended standards where needed and influencing present standards and necessary changes and to provide end user satisfaction and quality of service.	Strong data networking & telecommunications influence.	Vancouver, Canada	P: 604 527-1004
NAB	National Association of Broadcasters	Corporate	Serves and represents radio and television broadcast stations and networks, including their development and use of new technologies. Co-sponsors NAB Multimedia World Conference with IMA.	Represents television broadcast station owners' interests; strong policy and engineering history	Washington, DC	P: 202 429-4300 F: 202 429-5406

Name	Full Name	Principal Membership	Description/Mission	Comments	Location	Contact Information
NAMM	National Association of Music Merchants	Corporate	NAMM serves the industry by providing professional development programs; organizing trade shows and coordinating market development efforts aimed at unifying and strengthening the music products industry and increasing the number of active music makers.	Serves musical instrument dealers and manufacturers.	Carlsbad, CA	P: 619 438-8001
NCGA	National Computer Graphics Association	Corporate & Individual	A non-profit organization of individuals and major corporations dedicated to developing and promoting the computer graphics industry and to improving computer graphics applications in business, industry, government, science, and the arts.	Strong CAD/CAM and industrial design focus and history	Fairfax, VA	P: 703 698-9600 F: 703 560-2752
NCTA	National Cable Television Association	Corporate	The NCTA is the cable industry's major trade association. Its primary mission is to provide its members with strong national presence by providing a single, unified voice on issues affecting the cable industry.	Serves the interests of cable operators; strong legal and policy focus.	Washington, DC	202 775-3550
NMAA	National Multimedia Association of America	Individual & Corporate	Goal is to improve it's members knowledge and abilities while promoting and developing new markets and uses for multimedia applications.	Established in 1994; member benefits aimed at individual members	College Park, MD	P: 800 214-9531

Name	Full Name	Principal Membership	Description/Mission	Comments	Location	Contact Information
SALT	Society for Applied Learning Technology	Individual & Corporate	SALT provides a means to enhance knowledge and job performance of an individual by participating in Society sponsored meetings, and through receiving Society publications.	Professional Society for educators and trainers with an interest in computer based learning	Warrenton, VA	P: 703 347-0055 F: 703 349-3169
SMPTE	Society of Motion Picture and Television Engineers	Individual & Corporate	Disseminates technical information and provides a forum for the standardization of equipment, materials, and practices used in the industry. Membership.	Highly technical professional society with focus on film and TV technologies	White Plains, NY	P: 914 761-1100 F: 914 761-3115
SPA	Software Publishers Association	Corporate	International trade association for the personal computer software industry.	Focused on PC-based software companies with strong anti-piracy activities	Washington, DC	P: 202 452-1600 F: 202 223-8756
USTA	U.S. Telephone Association	Corporate	Lobbyist for local telephone services	US telecommunications companies	Washington, DC	P: 202 326-7300
VESA	Video Electronics Standards Association	Corporate	VESA sets and supports industry-wide video graphics standards for the benefit of end-users. Aims to reduce incompatibility among graphics boards, monitors, and software and, more recently, TV set-top devices (VOST)	Focuses primarily on PC related hardware standards (historically on graphics hardware).	San Jose, CA	P: 408 435-0333 F: 408 435 8225

About the Contributors

HARRY L. BOSCO is currently the ATM Platform Organization Vice President in AT&T Network Systems. Mr. Bosco began his career at AT&T in 1965 in the Data Communications area of Bell Labs. He was involved in the software and hardware development of the toll and local digital switching products from 1972 through 1983 when he became a laboratory director for Operational Support Systems. Mr. Bosco has since held positions directing new medical ventures in AT&T, leading the Data Networking Division. In 1992 he assumed his new position.

Mr. Bosco received an ASEE from Pennsylvania State University, BSEE from Monmouth College, and MSEE from Brooklyn Polytechnic Institute of New York.

GARY BRINCK has been in the computer systems development business for 30 years, 28 of them with IBM. During that time he has been a software developer, analyst, system architect, project manager and product strategist, covering projects ranging from the operating system for the S/380 mainframe to minicomputers and workstations to personal computers.

As a member of the development group for the Multimedia extensions to the OS/2 operating system, Mr. Brinck represented IBM in an effort to formulate practical industry standards through the Interactive Multimedia Association. Subsequently, he became Program Manager responsible for all software for the IBM TouchMobile system, a handheld computer designed to meet the needs of local motor freight carriers and delivery route personnel.

Retiring from IBM in 1993, Mr. Brinck now works as an independent software and systems consultant. Gary lives with his wife Nancy in the Ocala National Forest near the north Florida city of Ocala.

DR. WALTER S. CICIORA is a consultant in Cable, Consumer Electronics, and Telecommunications. Most recently he was Vice President of Technology at Time Warner. Walt joined Time Warner in 1982 after being with Zenith since 1965. He holds nine patents.

Walt currently serves on the Montreux Television Symposium Executive Committee and is a member of the board of directors at the Society of Cable Television Engineers, SCTE. He has served as chairman of: Technical Advisory Committee of CableLabs, National Cable Television Association (NCTA) Engineering Committee and the IEEE International Conference on Consumer Electronics. He is also a past president of the IEEE Consumer Electronics Society.

Walt is a Fellow of the IEEE, a Fellow of the Society of Motion Picture and Television Engineers, and Senior member of the Society of Cable Television Engineers. He received the 1987 NCTA Vanguard Award for Science and Technology and was named Man of the Year in 1990 and 1993 by CED magazine. Walt has a Ph.D. in Electrical Engineering from Illinois Institute of Technology and an MBA from the University of Chicago. He also received his BSEE and MSEE from IIT, and has taught Electrical Engineering in the evening division of IIT for seven years.

PHIL CLARKE is an MCI Strategic Global Services agent reselling enhanced data services ranging from Electronic Data Interchange (EDI) to Frame Relay, ATM, ISDN, and X.25. Phil is currently working on a nationwide network application called the Integrated Tribal Network (ITN) for a Native American owned company called Diversified Business Technologies, headquartered in Albuquerque, NM. Phil has spent the last 8 years working in the advanced messaging divisions of Western Union, AT&T, and MCI. He occasionally speaks and gives seminars on various messaging topics at conventions across the country.

Mr. Clarke has worked extensively on messaging applications for multinational companies and global organizations such as the World Bank, the International Monetary Fund, the OECD, Mack Trucks, Scott Paper, and Dupont. He also designed, programmed, and installed a communications switch at Hickam Air Force Base for the Emergency Communications Network Pacific group of the American Red Cross.

PHILIP V.W. DODDS was named Executive Director of the Interactive Multimedia Association (IMA) in January 1993, after serving as Managing Director and as the IMA Compatibility Project Director. Mr. Dodds also served as chairman of the Executive Committee for the National Association of Broadcasters (NAB) Multimedia World Conference and Exhibition.

In 1983, Mr. Dodds founded Visage, Inc., in Framingham, MA, a developer and manufacturer of multimedia products, including interactive video and digital audio products. Previously, Mr. Dodds was Director of Engineering for CBS, Inc. (Music Division), and Vice President of Engineering for ARP Instruments, Inc. ARP was a leading manufacturer of professional music instruments and produced the first electronic keyboard that connected to Apple and IBM PC computers, a forerunner of today's MIDI standard.

JIM GREEN is a Senior Program Manager in Microsoft's Advanced Consumer Technology Group. He is currently working on operating system support for multimedia and interactive television. He joined Microsoft in June of 1992 after completing work on the Audio Video Kernel (AVK) at Intel's DVI operation in Princeton, N.J. AVK is the multimedia systems software for Intel's ActionMedia™ products for Windows and OS/2. He has also been active in the Interactive Multimedia Association as chairman of the Architectural Technical Work Group. He has written several articles on multimedia for publications such as the Communications of the ACM, the IEEE Compcon Proceedings and Dr. Dobb's Journal, and has been a speaker or panelist a numerous conferences including Siggraph and NAB Multimedia World.

MONTY C. MCGRAW is a Program Manager at Compaq Computer, where he manages new consumer PC product development. He has been involved in personal computer technology and product development more than seventeen years, including the first 386 PC, and has six PC architecture patents.

During Mr. McGraw's tenure as chairman, the Interactive Multimedia Association Digital Audio Technical Working Group (IMA DATWG) published "Proposal for Standardized Digital Audio Interchange Formats" in May 1992. Mr. McGraw received his M.S. in Systems Management from the University of Southern California in 1980 and a B.S. in Electrical Engineering from Texas A&M University in 1974. He is a member of IEEE Computer Society and ACM.

Mitsubishi Electronics America is a wholly owned subsidiary of Mitsubishi Electric Corporation, a leading worldwide manufacturer of electrical and electronics products. Mitsubishi Electronics America employs over 4,000 people in research and development, manufacturing, sales, and service operations throughout the United States.

The North American Multimedia Business Center, established in September 1994, has a worldwide responsibility for advanced product planning and business development related to multimedia technology and its applications. The Center works with organizations throughout Mitsubishi Electric, as well as external strategic partners, to pursue opportunities that it identifies and initiates. As Vice President–Deputy General Manager, Mr. Mirowitz is responsible for all technical aspects of the Center's activities.

Previously, as Vice President–Advanced Product Planning, Mr. Mirowitz was responsible for U.S. market technical planning and business development for computers, computer peripherals, and electronic components. Mr. Mirowitz has managed several business and product startups in these markets for Mitsubishi Electronics America since joining the company in 1982. He has also worked at Western Digital Corporation as General Manager of the Imaging Business Unit, and at Touche Ross & Co. and McDonnell Douglas. Mr.

Mirowitz is a 1973 graduate of Dartmouth College and received his MBA in 1977 from the Wharton School of the University of Pennsylvania. He lives in Newport Beach, California.

MICHAEL C. RAU is Senior Vice President, Science & Technology, for the National Association of Broadcasters in Washington, D.C. NAB is a nonprofit trade association representing the interests of commercial radio and television broadcast stations and networks. He is responsible for management of the technology affairs of NAB, including service to NAB members, representation on industry committees, technology development and standards, and conference planning.

Mr. Rau graduated in 1978 from Clarkson University, Potsdam, New York, with a Bachelor of Science Degree in Physics. He then spent two and half years with his family's company, Rau Radio Stations, Inc., working in production and engineering. When the stations were sold, Mr. Rau briefly worked with engineering consultants Jules Cohen & Associates, Washington, D.C., and then accepted an offer from NAB in June of 1981. In December 1987, he was promoted to Vice President, Science & Technology, and then to Senior Vice President in April of 1990. In addition to a strong technical background, Mr. Rau put himself through law school at night. He received a Juris Doctor degree from Washington's Catholic University of America in May 1988, and is a member of the District of Columbia Bar.

Index

Corporation for Open Systems
    International (COS) 277, 300
CP/M 155
CRT 96
CSA 117
Customer Premise Equipment (CPE)
    66, 67, 71, 236
cyberspace 183

D
DAT 101
data
    encryption 147
    flow 220
    standards 230
    theft 147
    transmission 285
    types 186
DAVIC 106, 300
David Sarnoff Research Center 42
DCC 82, 101
DDD 260, 264
DDN 287
decompression 163, 176
Defense Data Network (DDN) 287
Department of Commerce 116
Department of Defense (DOD) 284
desktop conferencing system 201
desktop video 202
developer tools 180
digital audio 164
Digital Audio Video Council 300
digital audiotape (DAT) 101
Digital Audio-Video Industry
    Council (DAVIC) 106
digital broadcasting 30, 31
digital compact cassette (DCC) 101
digital multimedia 4
digital signal processors (DSPs) 98
digital standards 105
digital storage media command and
    control 300

digital television 113, 123
digital transmission 48
digital video 168
Digital Video Interactive 182
Direct-Distance Dialing (DDD) 260,
    264
distribution systems 266
divestiture plan 291
DMA 177
DOD 284
DPU 63
DSM-CC 300
DSPs 98
DVI 226
Dynabook 153

E
EDI 266
Edison, Thomas 80, 254
EFF 301
EGA 215
EGA (Enhanced Graphics Adapter
    board) 156
EIA 69, 89, 103, 105, 301
EISA 173
EISA (Extended ISA) 157
Electronic Data Interchange (EDI)
    266
Electronic Frontier Foundation 301
Electronic Industries Association 54,
    69, 301
electronic money 114
Electronics Industry Association
    (EIA) 84, 103
EMF 301
encryption 285
enterprise-wide computing 12
entertainment 125
error detection 285
European manufacturers 86
European Multimedia Forum 301

PAL 39, 41, 105
Paley, William S., of CBS 80
PBX 263
PCI 173
PCMCIA cards 136
PCS (Personal Communication Services) 144
PDA 31, 131
Permanent Virtual Circuit (PVC) 241
personal digital assistant (PDA) 31, 131
personal intelligent communicators (PICs) 31
personal-assist products 125
Philips 104, 112, 165
    Compact Disc Interactive (CD-I) 159, 174
    See also CD-I
Photo-CD 160, 173, 174
PICs 31
Pioneer Electronics 85
plain old telephone service (POTS) 257, 263
power management 137
PowerPC 175
presentation graphics 192
price caps 293
pricing practices 116
privacy 62, 67, 68, 75, 147
private-line service 263
Prodigy 260
programmed instruction 182
PS/2 series 157
PSNs 267
Public Television System 35
publishing formats 153, 172
PVC 241

Q
Qualcomm 142
quantizing 186
QuickTime 174   See also Apple QuickTime

R
RACONET 141
radio 27, 253
Radio Corporation of America (RCA) 27, 28, 80, 85
radio frequency (RF) 79, 83
    local-area networks (RF LANs) 143
    wide-area networks (RF WANs) 141
    spectrum 50
Radio Manufacturers Association 28
radiotelegraphy 79
RAM Mobile Data 142
ratings 34
RBOCs 267
real-time multitasking 177
real-time scheduling 212
recording format standards 101
Regional Bell Operating Companies (RBOCs) 267
repeaters 272
resource management 210
retailers 90
retailing channels 90
return call 259
RIFF (Resource Interchange File Format) 217
right of way 275, 270
RISC (reduced instruction set) 175
Rockwell International 142

S
SALT 306
sampling 186
Sarnoff, David 28, 80
satellite 253, 273
Satellite networks 142
SCCs 267
schedulers 213
Script-based tools 197
Scripting 171
scripting languages 122

SDH 238
SECAM 39, 41, 103, 105
security 62, 67, 70
Sega 134, 138, 203
set-top box 61, 63, 66
signal interdiction 62
smart card 70
smart multiplexer 282
SMDS 278, 279
SMDS/CBDS 240
SMPTE 103, 306
SMR (Specialized Mobile Radio) 141
SNA 286
Society for Applied Learning
    Technology 306
Society of Motion Picture and
    Television Engineers 306
Society of Motion Picture and
    Television Engineers (SMPTE)
    103
software compatibility 175
Software Publishers Association
    306
software standardization 229
software-based decompression 177
SONET 238, 280
Sony 83, 85, 104, 112, 165
SPA 306
Specialized Common Carriers (SCCs)
    260, 267
speed calling 259
speed dialing 259
Sprint 256, 260
standard formats 230
standardization 62
standardized interfaces 173
standards 38, 102
state regulations 293
Station Ownership Regulations 50
subscriber privacy 62
SVGA 215
switched cable architectures 73
switched data 277

switched multimegabit data services
    (SMDS) 278
switching 281
synchronization 210, 214
Synchronous Digital Hierarchy
    (SDH) 238
Synchronous Optical Network
    (SONET) 238
syndicator 36
system software 180

T
T1 155, 263, 277
Tandy Video Information System
    (VIS) 160, 174
TBS 111
TCP/IP 284, 287
telecommunications 15
teleconferencing 253, 264
tele-education 248
telegraph 27
Telephone Consumer Protection Act
    294
teletraining 265
television 6
The Institute of Electrical and
    Electronics Engineers 302
Thomson Consumer Electronics 42
Time Warner Inc. 296
title 196
tone block 259
transducers 94
transition effects 193
transponder 273
transport 285
TRS-80 155
TV
    microwave booster stations 50
    pickup stations 50
    relay stations 50
    STL stations 50
    translator relay stations 50
    Tuner 63